POLITICS AND ECONOMICS OF THE MIDDLE EAST

ENVIRONMENTAL STRESS IN PAKISTAN AND U.S. INTERESTS

POLITICS AND ECONOMICS OF THE MIDDLE EAST

Additional books in this series can be found on Nova's website under the Series tab.

Additional E-books in this series can be found on Nova's website under the E-books tab.

ENVIRONMENTAL REMEDIATION TECHNOLOGIES, REGULATIONS AND SAFETY

Additional books in this series can be found on Nova's website under the Series tab.

Additional E-books in this series can be found on Nova's website under the E-books tab.

POLITICS AND ECONOMICS OF THE MIDDLE EAST

ENVIRONMENTAL STRESS IN PAKISTAN AND U.S. INTERESTS

OMAR SIDDIQUI
EDITOR

Nova Science Publishers, Inc.
New York

For permission to use material from this book please contact us:
Telephone 631-231-7269; Fax 631-231-8175
Web Site: http://www.novapublishers.com

Library of Congress Cataloging-in-Publication Data

Environmental stress in Pakistan and U.S. interests / editors, Omar Siddiqui.
p. cm.
Includes index.
ISBN 978-1-61470-224-5 (softcover)
1. Pakistan--Politics and government--1988- 2. Environmental degradation--Political aspects--Pakistan. 3. Political stability--Pakistan. 4. National security--Pakistan. 5. Security, International--South Asia. 6. Pakistan--Foreign relations--United States. 7. United States--Foreign relations--Pakistan. I. Siddiqui, Omar.
JQ629.A58E68 2011
363.7095491--dc23

2011022169

Published by Nova Science Publishers, Inc. † New York

CONTENTS

PREFACE

This book focuses on the nexus between security and environmental concerns in Pakistan that have the potential to affect American security and foreign policy interests. Environmental concerns include, but are not limited to, water and food scarcity, natural disasters and the effects of climate change. Environmental stresses, when combined with the other socio-economic and political stresses on Pakistan, have the potential to further weaken an already weak Pakistani state. Such a scenario would make it more difficult to achieve the U.S. goal of neutralizing anti-Western terrorists in Pakistan. Some analysts argue that disagreements over water could also exacerbate existing tensions between India and Pakistan. Given the importance of this region to U.S. interests for many reasons, these issues may be of increasing concern to Congress in the years ahead.

Chapter 1- This report focuses on the nexus between security and environmental concerns in Pakistan that have the potential to affect American security and foreign policy interests. Environmental concerns include, but are not limited to, water and food scarcity, natural disasters, and the effects of climate change. Environmental stresses, when combined with the other socio-economic and political stresses on Pakistan, have the potential to further weaken an already weak Pakistani state. Such a scenario would make it more difficult to achieve the U.S. goal of neutralizing anti-Western terrorists in Pakistan. Some analysts argue that disagreements over water could also exacerbate existing tensions between India and Pakistan. Given the importance of this region to U.S. interests for many reasons, the report identifies an issue that may be of increasing concern for Congress in the years ahead.

Chapter 2- Pakistan experienced a catastrophic natural disaster that has precipitated a humanitarian crisis of major proportions. Widespread flooding affected about 20 million Pakistanis and inundated an area the size of Florida within the country. Congressional interest in the flooding stems from the significant humanitarian and economic implications for the country, and the security implications for U.S. interests in the region. The World Bank and Asian Development Bank have estimated that the flooding has caused $9.7 billion in damages. While this figure might still be preliminary, it is almost certain that the negative effects of this crisis will be felt for many years to come.

In: Environmental Stress in Pakistan …
Editor: Omar Siddiqui
ISBN: 978-1-61470-224-5

Chapter 1

SECURITY AND THE ENVIRONMENT IN PAKISTAN*

Bruce Vaughn, Nicole T. Carter, Pervaze A. Sheikh and Renée Johnson

SUMMARY

This report focuses on the nexus between security and environmental concerns in Pakistan that have the potential to affect American security and foreign policy interests. Environmental concerns include, but are not limited to, water and food scarcity, natural disasters, and the effects of climate change. Environmental stresses, when combined with the other socio-economic and political stresses on Pakistan, have the potential to further weaken an already weak Pakistani state. Such a scenario would make it more difficult to achieve the U.S. goal of neutralizing anti-Western terrorists in Pakistan. Some analysts argue that disagreements over water could also exacerbate existing tensions between India and Pakistan. Given the importance of this region to U.S. interests for many reasons, the report identifies an issue that may be of increasing concern for Congress in the years ahead.

The report examines the potentially destabilizing effect that, when combined with Pakistan's demographic trends and limited economic

---

* This is an edited, reformatted and augmented version of a Congressional Research Service publication, CRS Report for Congress R41358, from www.crs.gov, dated August 3, 2010.

development, water scarcity, limited arable land, and food security may have on an already radicalized internal and destabilized international political-security environment. The report considers the especially important hypothesis that the combination of these factors could contribute to Pakistan's decline as a fully functioning state, creating new, or expanding existing, largely ungoverned areas. The creation, or expansion, of ungoverned areas, or areas of limited control by the government of Pakistan, is viewed as not in U.S. strategic interests given the recent history of such areas being used by the Taliban, Al Qaeda, and other terrorist groups as a base for operations against U.S. interests in the region. In this sense, environmental stress is viewed as a potential "threat multiplier" to existing sources of conflict.

Environmental factors could also expand the ranks of the dispossessed in Pakistan, which could lead to greater recruitment for radical Islamist groups operating in Pakistan or Afghanistan. Larger numbers of dispossessed people in Pakistan could also destabilize the current political regime. This could add pressure on the Pakistani political system and possibly add impetus to a return to military rule or a more bellicose posture towards India. This issue has added significant importance to regional security and American interests in Afghanistan.

The potential for environmental factors to stoke conflict between the nuclear armed states of India and Pakistan is also a concern. These two historical enemies have repeatedly fought across their international frontier and have yet to resolve their territorial dispute over Kashmir. Further, a longstanding dispute over cross-border water resource sharing between India and Pakistan has resurfaced, possibly exacerbating existing tensions between the two states. Should the two countries wish, however, this dispute also offers a renewed opportunity for cooperation, as has been seen in past negotiations.

Preliminary findings by experts seem to indicate that existing environmental problems in Pakistan are sufficiently significant to warrant a close watch, especially when combined with Pakistan's limited resilience due to mounting demographic stresses, internal political instability, security challenges, and limited economic resources.

INTRODUCTION

This report explores the nexus between the environment and security in Pakistan in order to assess how environmental stress in Pakistan can lead to security issues that affect American security and foreign policy interests in the region. Environmental crises such as water scarcity, soil depletion, and natural disasters can intensify conflict or stress within a country and potentially contribute to national security issues.[1] Climate change can fuel these crises and exacerbate them, leading some analysts to characterize it as a potential threat to national security. Others temper this characterization by noting that the potential effects of climate change could act as a *threat multiplier* to national security. In other words, the effects of climate change might exacerbate existing threats to national security such as weak governance, poverty, and armed insurgents.

The security environment within and around Pakistan is of significant concern to the United States. During the Cold War, the United States worked with Pakistan to meet common security challenges in Afghanistan and the region. Today's security threats to the United States from Pakistani territory and the region emanate from sub-national radical Islamist groups that oppose the United States and its allies. Pakistan's limited and tenuous control of much of its territory, the growing strength of radical Islamist groups in Pakistan, the poor state of Pakistan's economy, and ongoing political turmoil among Pakistani political elites all undermine Pakistan's ability to effectively control radical Islamist elements and to ameliorate growing environmental, economic, and other stresses. Pakistan's status as a nuclear weapon state, its traditional enmity with India, and proximity to Afghanistan all heighten its importance to U.S. strategic interests.

The environment and natural resources within Pakistan are also under stress. Once abundant water supplies are dwindling due to an increase in demand and drought. This is, in part, affecting crop production and reducing the area of arable land in the country. Pakistan has one of the highest deforestation rates in the world and currently has approximately 4% of its original forested area intact. Loss of forests has led to soil erosion and altered ecosystems in the region. The projected effects of climate change in Pakistan range from increased periods of drought and low water supplies to sea-level rise and associated saltwater intrusion. Many of these environmental stressors have potential socio-economic and political consequences for Pakistan. For example, drought can stress water supplies along the Indus River and potentially exacerbate tensions between Pakistan and India. Drought can further lead to reduced hydropower supplies and catalyze protests in areas experiencing rolling blackouts, and of course contribute to economic stresses in Pakistan's agricultural regions, where the majority of Pakistanis live.[2]

Existing environmental stress and potential future stress from climate change in Pakistan may undermine American interests in the region by leading to further socio-political instability in Pakistan. Alone, environmental stress might not become a geopolitical concern, but when added to existing political and socio-economic stresses, it has the potential to be geopolitically important due to the instability that it could create. Congress could be interested in this connection because of its role in authorizing and appropriating funds for Pakistan.[3]

CONTEXT FOR ENVIRONMENT AND NATIONAL SECURITY

Theoretical and empirical work on the relationship between the environment and security, which can help us better understand how environmental stress may affect Pakistan, has been expanding for decades. This report touches briefly on just a few of these sources to provide a sense of this growing body of knowledge and set a theoretical context for our analysis of Pakistan's water resources and food security.

J.R. McNeill, in an historical analysis of natural shocks to societies, has observed that such shocks have proved to be both unifying and divisive, that social conflict has been routine during and after natural disasters, that minorities or foreign groups are often blamed, and that government authorities have often been the target of popular wrath for failing to minimize or prevent damage. McNeill further points out that "Societies with little in the way of safety net ... easily succumb to banditry, ethnic and religious violence, and even outright civil war under the stress of acute drought. Restraint and civility can quickly perish when people are confronted with imperious necessity."[4]

Given this perspective, the fragility of the ecosystem and political structure of Pakistan may mean that it is more likely to be depleted or degraded due to environmental stress than more resilient states subject to the same pressures. Pakistan is subject to both demand-induced scarcity and structural scarcity.[5] Demand-induced scarcity stems from Pakistan's growing population and per capita use of resources. Structural scarcity stems from the inequitable distribution and use of natural resources in Pakistan, stemming from social inequality.[6]

Thomas Homer Dixon describes the dynamic of how environmental factors interact in a security context:

> If market failure, social friction, and capital availability prevent a society from supplying the amount of ingenuity that it needs to adapt to environmental scarcity, then five kinds of social effect are likely: constrained agricultural productivity, constrained economic productivity, migration, social segmentation, and disruption of legitimate institutions.

> Environmental scarcity is not the sole or sufficient cause of these social effects. Contextual factors range from the nature of relations among ethnic groups to the state's degree of autonomy from outside pressure groups.[7]

One can see how the above statement may relate to Pakistan when considering the threat multipliers at play in Pakistan today.

In his book *Collapse: How Societies Choose to Fail or Succeed*, Jared Diamond found that societies do not collapse due "solely to environmental damage: there are always other contributing factors." Diamond identifies a five-point framework of such contributing factors including: environmental damage, climate change, hostile neighbors, friendly trade partners, and the society's response to its environmental problems. Diamond found that this fifth factor, the society's response to environmental factors, "always proves significant."[8]

As the potential effects of environmental degradation and climate change on Pakistan's and many other nations' security become better understood, there may be greater impetus for the United States to assess international climate change mitigation and adaptation frameworks. The combination of environmental degradation, anticipated future impacts of climate change, and increasing demographic pressure are likely to place significant stress on many developing nations across the globe in the years ahead. Security, however, is complex, and climate change is but one threat among many security threats. Moreover, the security concern (e.g., failed states, rise of terrorists groups and unfriendly regimes, global economic and financial security, humanitarian and human development disasters) varies depending on the nation at risk (e.g., Pakistan) and the perspective of those affected by the risk (e.g., United States).

Figure 1 is an illustration of how the physical impacts of climate change may lead to socioeconomic effects, which may contribute to a threat to security and stability.

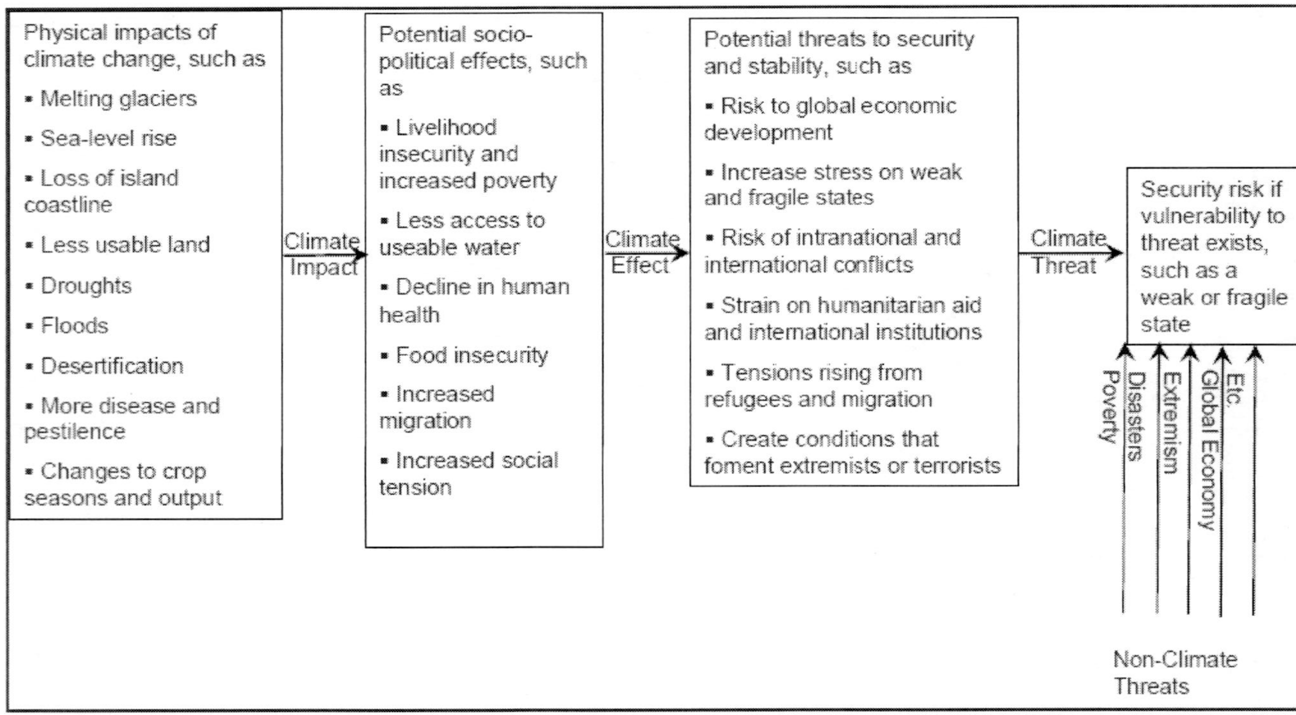

Source: CRS.

Notes: First two boxes were adapted from S. Smith and J. Vivekananda, A Climate of Conflict (International Alert, Nov. 2007), pp. 10-11, available at http://www.international-alert.org/pdf/A_Climate_Of_Conflict.pdf.

Figure 1. Illustrative Pathway of How Climate Change May Affect Security.

Figure 2 shows a map produced by the German Advisory Council on Global Change that overlays regional hot spots with anticipated climate change impacts in those areas. This demonstrates that the issues under consideration in this report on Pakistan are in various ways relevant to other parts of the globe as well.

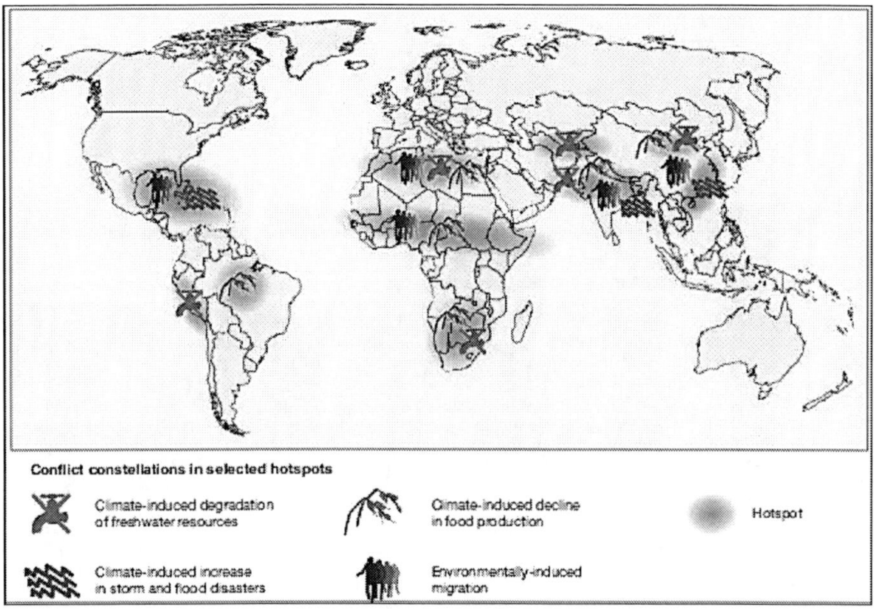

Source: R. Schubert, et al., Climate Change as a Security Risk (German Advisory Council on Global Change, May 2007), p. 4, available at http://www.wbgu.de/wbgu_jg2007_engl.pdf.

Figure 2. Select Security Hot Spots and Their Climate Threats.

WATER RESOURCES

Climate Change and Pakistan's Water Resources

Pakistan, which depends heavily on the Indus River system, is a water-stressed country. That is, most of the surface and groundwater resources are nearly fully exploited or are being used at an unsustainable rate. Water allocations among Pakistanis are highly inequitable.[9] Figure 3 provides data on the past and anticipated per capita water supply in Pakistan; the illustration shows the declining availability of water which affects both human health and livelihoods, and ecosystem health and the services provided by those ecosystems. Most Pakistani households do not have access to adequate potable or shallow water; 65% of households get their water from a pump.[10]

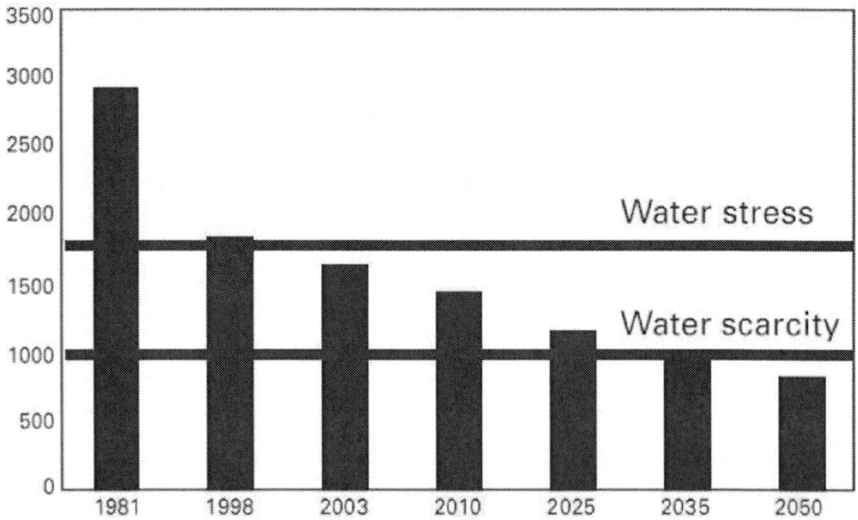

Source: World Bank 2005 data adapted by F. Khan, "Water, Governance, and Corruption in Pakistan," in *Running on Empty: Pakistan's Water Crisis* (Woodrow Wilson International Center for Scholars, 2009), p. 83.

Figure 3. Declining Per Capita Water Availability in Pakistan. (cubic meters per person annually).

Only a few decades ago, Pakistan was considered to have an abundance of water. The change results largely from the increased demand for water by expanded irrigation, population growth, less than optimal water use practices, and economic development. The near- and long-term effects of climate change on Pakistan's water-resources will further challenge the country's water sector, resources, and regional and ethnic tensions.

Water resource development for agriculture contributed to Pakistan's economic growth and rise as a regional food basket. Agriculture represents almost 97% of the water withdrawals in Pakistan. This is a very high percentage and leaves little for other uses. Pakistan now has the world's largest contiguous irrigated area, which is fed by surface water from the Indus River system and from groundwater wells. The irrigation network alone contributes to nearly a quarter of the country's gross domestic product and meets most of its food and fiber requirements. The livelihood of many families and the economy of the country, therefore, is sensitive to the availability of water and to reductions in availability that climate change may cause.

Figure 4. Map of Pakistan.

The growing imbalance between Pakistan's water supply and demand has led to shortages, regional competition, conflicts between stakeholders, and constraints on economic development. The tools available to address the imbalance range from trying to develop new supplies to improving the efficient use of existing supplies. Developing new sources through additional storage has been controversial. Improving water efficiency requires political will and significant investment. Determined efforts to strengthen institutional capacities in the water sector at all levels of government and to change behaviors of farmers have not been sufficient to effectively deal with the problem. The poor state of government finances limits the options available to the government.

Under climate change predictions, the eventual loss of glacier storage may significantly alter water available in the Indus River for use in Pakistan. Western Himalayan glaciers are projected to continue their retreat for the next half century, increasing Indus River flows and flooding. These initial increases in river flows will likely be followed by decreased river flows as the glacial melt decreases as the glaciers disappear.[11] A 2010 Dutch study found that melt water from the Himalaya accounts for 60% of the water in the Indus river. The study also found that projected temperature, rainfall, and snow projections would likely lead to a 8.4% decrease in upstream water flow into the Indus by 2050 which would threaten the food security of those dependant on the river for irrigation.[12] This is despite a projected increase in rainfall in the area. The change in the glacial melt is not the only water-related effect on Pakistan anticipated under climate change. Climate change has been projected to increase the variability of monsoons, decrease the predictability of precipitation, exacerbate water-stress in arid and semi-arid regions, and further salt-water intrusion of coastal freshwater supplies.[13]

AGRICULTURAL PRODUCTION AND FOOD SECURITY

Pakistan's agriculture sector is a major part of its economy, contributing to about 20%-30% of the country's Gross Domestic Product (GDP), with about 70% of the population dependent on agriculture.[14] Roughly 45% of the working population is employed in agriculture, forestry, and fishing. As illustrated by statistics reported by the Food and Agriculture Organization (FAO) of the United Nations (U.N.), however, expected water shortages and soil degradation for Pakistan, among other sectoral factors, may effectively constrain expansion of its agricultural resources and production.

Pakistan's agricultural production includes both food and cash crops. Major food crops are cereal grains (mostly wheat and rice, and also barley, corn, millet, and sorghum) and milk products (buffalo, cow, and goat). Other commodities include other products such as fruits, vegetables, cattle, mutton, and eggs. Most livestock (60%-90%, depending on the area) are pasture- or rangeland-fed and are not generally fed cereal grains which are needed for human consumption.[15] Most food production is centered in the Punjab and Sindh provinces, which are the center of the country's wheat and rice production. (See Figure 4.) Mixed crop production areas are more prevalent in the Northwest Frontier Province (NWFP) areas; Baluchistan is characterized by mostly arid/sparse production, along with limited pasture-based livestock production.

Pakistan's major agricultural cash crops are cotton, sugarcane, and tobacco. Cotton serves as an important input into Pakistan's textile and apparel industry, and competes for available land, waters, and other resources and production inputs.[16]

Declining Agricultural Production

In the last decades of the 20th Century, Pakistan's agricultural production grew sharply, increasing by an average of 4% per year, in large part due to the use of high-yielding crops, increased government prices for crops, and subsidies for irrigation water, fertilizer, and other inputs.[17] In recent years, however, agricultural production has been declining. Among the many reasons cited for this decline are:[18]

- shortage of irrigation water and deterioration in the irrigation network, including severe drought conditions in 2000-2001;
- soil degradation from fertilizers and other chemical inputs, and low efficiency from most farm inputs;
- low technology use, and lack of farmer knowledge of operating and technological characteristics within the agricultural sector;
- large differences in yields among provinces;
- uncertain land tenure;
- poor government investment in agricultural research and marketing;
- domination of large landowners and other vested interests, including price control by cartels;
- lack of crop diversification and focus on wheat production, particularly by some donor organizations; and
- poor governance and corruption.

The U.N. reports that although Pakistan was a net exporter of wheat and other cereal grains in the 1980s and 1990s, Pakistan became a large net importer of food products during the 2001-2003 period.[19]

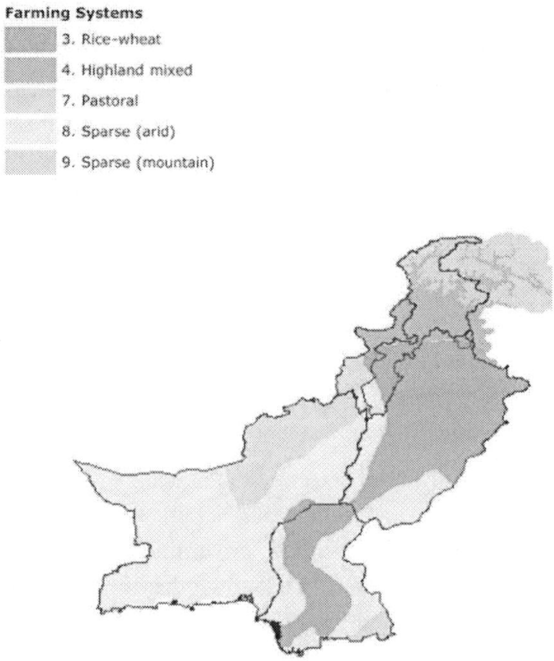

Source: U.N. FAO, « Pakistan, Agriculture Sector, » http://www.fao.org/countryprofiles/index.asp?lang=en& iso3=PAK&subj=4.
Notes: FAO study, "Farming Systems and Poverty: Improving Farmers Livelihoods in a Changing World," 2001, http://www.fao.org/farmingsystems.

Figure 5. Pakistan Farming Systems.

Land Use and Irrigation

Less than 30% of Pakistan's total land area (about 21.3 million hectares) is under cultivation and can be intensively farmed.[20] Most arable land and land under cultivation is in the densely populated Punjab and Sindh provinces. This coincides with data showing that these same two provinces account for nearly 90% of Pakistan's irrigated agriculture.

Nearly all available water in Pakistan is used for agriculture. As mentioned above, Pakistan's agriculture sectors rely heavily on irrigation, given its generally arid climate and low soil moisture. About 10% of Pakistan's cultivated area is classified as "semi-arid," with the majority of cultivated land classified as either "arid" (48% of total) or "hyper-arid" (37%).[21] By major province/region, about 60% of the arable land in both Baluchistan and NWFP is considered to consist of dry areas, with about 50% of the arable land in both Sindh and Pakistan's Federally Administered Tribal Areas (FATA) bordering Afghanistan classified as dry. The share of dry areas in Punjab is lower at 23% of all cultivated land. The major factor attributed to land degradation in Pakistan is low organic matter in soils, estimated at less than 1%.[22] Other factors include soil erosion, underuse of organic fertilizers and overuse of synthetic chemicals, salinity and sodicity, and waterlogging.[23] Overly extensive cultivation in rainfed areas is common.

Pakistan has one of the world's largest irrigation systems. Irrigation covers an estimated 83% of all cultivated land in the country. About 72% of irrigated agriculture is centered in the Punjab region, the principal food production region in the country, with about 10.3 million hectares under irrigation. The Sindh province accounts for another 2.5 million hectares (17%) of irrigated agriculture. Comparable estimates for Baluchistan and the North-West Frontier Province (NWFP) are about 700,000 hectares under irrigation, or about 5% each of Pakistan's total irrigated area. [24]

Table 1.Distribution of Irrigated Area, Pakistan by Federal State Territory

Federal state territory	Irrigated area (hectacres)	Irrigation (percent)
Baluchistan	767,120	5%
North-West Frontier Province (NWFP)	719,152	5%
Punjab	10,325,678	72%
Sindh	2,515,050	17%
Northern Territories	90,464	1%
Pakistan, total	14,417,164	100%

Source: U.N. FAO, Aquastat, "Global Map of Irrigation," http://www.fao.org/nr/water/aquastat/irrigationmap/pk/ index.stm. Estimates have origin in 1990 Census.

Notes: Areas equipped with irrigation infrastructure. Does not include other irrigated or rainfed areas, which total another roughly 8 million hectares.

Food Security and Nutrition

The UN notes that fluctuations in the availability of food is among the "factors responsible for the micronutrient deficiency disorders observed in Pakistan."[25] For the 2001-2003 period, the U.N. reports that energy supplied from wheat and other cereals accounted for about 38% of total dietary intake, along with milk products (11%) and sugar products (10%).[26] Still, more than 20% of Pakistan's population is reported to suffer from food deprivation.[27]

The U.N. reports that the nutritional status of Pakistani children under five years of age is "extremely poor."[28] At a national level almost 40% of children under five years are underweight; about 37% of children are affected by stunting and about 13% by wasting.[29] Moreover, the U.N. reports that there are significant variations in malnutrition rates among provinces. For example, "The prevalence of stunting appears to be associated with the overall level of development of the provinces, being lowest in Punjab and highest in Baluchistan, the least developed province."[30] Baluchistan, NWFP, and the FATA areas, along with areas with large numbers of internally displaced persons (IDPs), are among the most food insecure in Pakistan.[31]

Natural Disasters

Pakistan is prone to certain types of natural disasters with significant impacts, especially earthquakes and floods. Pakistan has experienced major earthquakes that have caused considerable fatalities and damage to critical infrastructure. The last major earthquake in Pakistan with significant consequences was in northern Pakistan in October 2005.[32] Over 73,000 people died as a result of the earthquake and over 5.0 million were displaced.[33] This disaster created issues related to food security, health and disease, water and sanitation, and infrastructure. It also had a large economic toll, causing some to estimate that recovery could cost over $5.0 billion.

Pakistan is also subjected to flooding during the monsoon season, when flooding has the potential to displace tens of thousands of people, damage infrastructure, and destroy croplands. In relation to climate change, flooding patterns might follow changes in monsoon seasons. A World Bank study has stated that between 1990 and 2008 natural disasters killed 60,000 and affected 750 million people in South Asia with $45 billion in damages.[34] In July and August of 2010, Pakistan experienced what have been described as the worst floods in the country's history. These floods reportedly killed over 1,100 and devastated large parts of the Swat Valley where the government of Pakistan is seeking to reassert its control after displacing Islamist militants. A Pew Research poll has found that only 17% of Pakistanis hold a favorable view of the United States.[35] In August 2010, the United States announced $10 million in assistance to aid those affected by the flooding. Such assistance may help improve Pakistanis' perceptions of the United States. Melting glaciers might lead to glacial lake outburst floods,[36] which can affect communities and settlements downstream. A burst can discharge millions of cubic meters of water and debris in a few hours into downstream communities. There are over 2,500 glacial lakes in the Himalayan region of Pakistan; however, just a small fraction are considered dangerous.[37]

Natural disasters often lead to displaced populations and damaged infrastructure in countries such as Pakistan. The response to these events can stress governments and lead to or exacerbate internal conflicts. These events can also provide radical groups an opportunity to help displaced people and recruit new members, when government responses are weak. The promptness and efficiency of militants in providing assistance to some of those affected by the 2005 earthquake in Azad Kashmir and the Northwest Frontier Province of Pakistan earned them a degree of legitimacy within communities affected by the quake.[38]

Environmental Stress as Threat Multiplier to Security in Pakistan

While much attention has been devoted to environmental stress that is anticipated due to climate change, existing environmental concerns already stress Pakistan. If these conditions worsen, they could act as a threat multiplier in combination with other threats to the state such as expanding population and food insecurity, strife among political-military elites, poor economic resources, Islamist extremism, secessionism, inter-provincial competition for resources, and/or cross border conflict with Afghanistan or India. Such stress could further weaken the state to the extent that it could fall to pressure by militants. Such a situation could also lead to a return to military rule.

Demographics and Food Insecurity

High population growth and density, low economic development, and resource scarcity may lead to increased risk of conflict. Pakistan contains all of these factors to certain degrees. Environmental stress can exacerbate these existing problems by increasing resource scarcity (i.e., limiting food production). Food production in Pakistan is a key factor for development because of the limited carrying capacity of arable land and state resources relative to Pakistan's population. Food production is not keeping pace with Pakistan's steadily increasing population. Pakistan's population almost quadrupled from 1960 to 2010, and is expected to grow by another 66 million in the next 15 years, thereby reaching 335 million by 2050.[39]

A combination of increasing population size, limited arable land, and poverty put Pakistan at extreme risk of food insecurity, according to some studies. For example, in 2009, Pakistan ranked 11th on a list of the world's Food Security Risk Index which placed it in the category of states at extreme risk.[40] The number of people in Pakistan considered "insecure" in terms of their food needs reportedly increased from 35 million to 45 million in 2009. Some reasons for this status are economic. For example, the price of wheat flour in Pakistan increased 95% from May 2007 to October 2009 while overall food inflation increased prices 28% in the year preceding October 2009. An estimated 85% of Pakistanis live on two dollars a day or less and spend at least 50% of their income on food. Pakistan produces approximately 20 million tons of wheat while consuming 23 million tons, leading to the necessity of wheat imports.

"Without food, people do one of three things: revolt, migrate, or starve. When governments can no longer provide food security, states fail."[41] From May 2007 to October 2009 (when wheat flour increased in price), riots in Pakistan reportedly led to the stealing of massive amounts of grain and flour. In January 2008, Pakistan deployed 6,000 soldiers to secure flour mills, food distribution centers, and to escort supply trucks.[42] The food security situation is considered to be the worst in Pakistan's northern and western border areas.[43] In Baluchistan 20 of 25 districts were considered food insecure while all of the Federally Administered Tribal Areas were considered food insecure.[44] These areas are the same areas that are subject to much of Pakistan's internal strife.

Refugees place added pressure on food security in Pakistan. Regional conflict in Afghanistan has made Pakistan one of the leading locations of refugees over the years. Millions of Afghans fled to Pakistan to avoid fighting during both the Soviet and American occupations of Afghanistan. Internal fighting between the Pakistan government and Taliban forces in northwest Pakistan in 2009 led to an estimated 3 million Internally Displaced Persons (IDPs).[45] Such dislocated populations lack the resources of more stable populations and can contribute to food scarcity and overtaxed infrastructure, among other things. Refugees have the effect of reducing the resilience of regional governments when facing environmental stress. Further, environmental stress, such as food scarcity and limited water supplies, exacerbates the effect of refugees and lead to protests, crime, and potentially recruiting by militants. Cross border migration appears to be less of an option for Pakistan's populations than in other states due to Pakistan's adversarial relationship with India and the ongoing conflict in Afghanistan.

Political-Military Elite Strife

Strife among the political-military elites of Pakistan is a key factor that prevents Pakistan's leadership from effectively deploying state resources for environmental and other issues. Changes made to the Constitution in 2010 are not thought likely to cure tri-partite conflict between the military, judiciary, and the government.[46] As a result, political infighting is expected to continue at the cost of effective governance.[47] Corruption remained a problem in 2009 as Pakistan ranked as the 42nd most corrupt of 180 states in a ranking by Transparency International.[48] Resource scarcity has also evidently led to elite's control of resources that has ecologically marginalized many of Pakistan's poor who are increasingly vulnerable.[49] Many of the urban poor are more vulnerable to price increases than the rural poor because they must buy all of their food. Pakistan is 36% urban with a 3% annual shift from rural to urban areas.[50]

Federal Minister of Investment Waquar Ahmed Khan announced in April 2009 that the government of Pakistan has plans to sell or lease one million acres of farmland to foreign investors who are thought likely to be from the Persian Gulf. The Pakistani Taliban asserted control over the Swat Valley, about 100 km north of Islamabad, in 2008 before the government reasserted its influence in the region. There are plans to raise a special security force to protect foreign investment in Pakistani farmland. A further question is the extent to which such investment will compete with Pakistani agriculture for water resources.[51] Food produced on such foreign-controlled farms may be exempt from export bans during times of shortage.

The Economy[52]

The effect of environmental stress on Pakistan can be, in part, related to the country's poor economic performance which limits Pakistan's resilience in dealing with environmental stress. Developed states with strong diversified economies are likely have greater resources to mitigate and adapt to environmental stress, such as climate change. Less developed states such as Pakistan might not effectively deal with environmental challenges. For example, a robust industrial sector could absorb excess workers from the agricultural sector should a decline in that area lead to its downsizing. Without alternative employment, potentially displaced agricultural workers could have difficulty finding work to buy food for their families.

The Pakistani economy has been hindered by decades of internal political conflict and low levels of foreign investment. Despite this, some gains have been made in poverty reduction, and GDP growth is forecasted to be 2.9% in 2010. Unemployment remains high at 15.2% in 2009. An estimated 24% of the country lives below the poverty line, and GDP per capita in purchasing power parity terms is $2,600 per person per year. Inflation is expected to remain high in 2010 at 11.3% even as it has dropped from 20.3% in 2008 and 13.6% in 2009. [53] Pakistan received emergency funds from the International Monetary Fund (IMF) in late 2008 to help stabilize its economy. While the economy stabilized somewhat in 2009, the IMF warned in February 2010 that challenges to the economy remained. Some believe that domestic factors will hinder government efforts to further improve economic stability. The government's resources will be further strained when it has to repay the IMF loans.

Environmental stress such as low agricultural productivity, declining water supplies, and natural disasters can exacerbate economic problems. Low agricultural productivity could lower GDP, increase unemployment, and potentially raise prices for staple foods. Low water supplies could affect agriculture and industry dependent on water, as well as, directly affect the populace. Lastly, natural disasters can exact a financial burden on the government and lead to displaced people without jobs. The sum impact of these environmental stresses would be to exacerbate the existing problems in the Pakistani economy. The degree of their effect is related to the intensity and timing of the environmental stress. A poor economy limits Pakistan's ability to deal with security threats as it limits the state's resource base that it can deploy to address the threat.

Islamist Extremism

Islamist extremism in Pakistan is a direct security threat to the United States, Pakistan, and other countries.[54] Environmental stress in Pakistan can contribute to the growth and sustenance of Islamic extremism in the region. One documented example connecting environmental stress and Islamic extremism was the major earthquake in Pakistan in 2005. It was reported that militant groups in Pakistan played an active role in providing aid after the earthquake. These actions were done to bolster their image and possibly gain recruits. Natural disasters leading to a weak government response has the potential to create discord among displaced populations. In some cases, it might empower fringe, militant groups that pose a security threat to the country; whereas in other cases, it might result in protests and disassociation with the ruling party.

Pakistan's Periphery

Key existent stresses on the Pakistani state stem from autonomous and secessionist tendencies in Pakistan's western borderlands, particularly in the Federally Administered tribal Areas (FATA), Northwest Frontier Province (NWFP) (recently renamed the Kyber Pakhtunkhwa), and Baluchistan. Pakistan's central government has never extended the same degree of control over these areas that it does over its core in Punjab and Sindh. This was also the case under the previous British colonial administration. A deteriorating environment due to increased stress on the land that could come about as a result of potential demographic pressure could further weaken central government control over these areas. It was reported in April 2010 that militants were still using sanctuaries in Pakistan's FATA to strike U.S. and NATO forces in Afghanistan.[55] This is in addition to Waziristan having connections to recent failed attempts to attack the United States and its interests in the region. Baluchistan is an area where long standing tensions with the centre in Islamabad overlap with an area of current water stress.

Inter-Provincial Competition for Water Resources

An ongoing dispute over the construction of the Kalabagh Dam on the Indus River in the Pakistani Province of the Punjab highlights intra-Pakistani tensions over water resource sharing. The Pakistani provinces of Sindh, Baluchistan, and the Northwest Frontier Province oppose the dam's construction while the Punjab supports it as a means of solving both energy and water shortage problems. Pakistan's Federal Minister for Water and Power Raja Pervaiz Ashraf has stated that Pakistan is desperately in need of establishing new dams in order to overcome Pakistan's power crisis and overcome load shedding.[56] The Lahore Chamber of Commerce and Industry has estimated that the dam would produce enough energy to obviate the need to import 20 million barrels of oil a year.[57]

Sindh, which is the lower riparian province, has been the most vocal opponent of the project for several reasons. Many in Sindh fear that the dam will irrigate land in the Punjab at the cost of irrigating land in Sindh. Sindh also fears that the dam will reduce Indus water flow to the Arabian Sea, which is essential in order to prevent salt water intrusion from destroying coastal mangrove stands and farm land. An estimated 85% of the Indus River Delta mangroves, which are important spawning grounds for fish and crustaceans, have been lost. Flow diversion, leading to increased salinity levels, as well as industrial pollution from Karachi and the local cutting of mangroves for fodder and fuel wood, is degrading the ecosystem.[58]

If current water use plans proceed in the lower riparian area of the Indus, Sindh may be placed in extreme jeopardy. According to one expert, "If the Vision-2025 is implemented as planned, the lower basin will receive practically no water. If the situation is not reversed soon and the storage and diversion upstream continues, it is not farfetched to see an impending ecological disaster in the lower Indus basin that would eventually create a famine-like situation in Sindh."[59] Salt water intrusion has been detected as far as 100 kilometers inland in Sindh. Reportedly hundreds of villages in the Badin and Thatta Districts have had to migrate to the interior of Pakistan as salt water intrusion has rendered their farmlands useless. Sea level rise associated with climate change is likely to exacerbate this already dire situation.[60]

Cross-Border Dispute with India over Kashmir and Water Resources

Pakistan and India are historical enemies that have fought three wars and had numerous periods of heightened tension over their disputed line of control in Kashmir since their independence from British colonial rule in 1947. Pakistan and India most recently fought over the line of control in Kashmir near Kargil in 1999 after Pakistani forces crossed the de facto line of control in a failed attempt to occupy Indian-held territory.

The 1960 Indus Water treaty allotted the water from the Ravi, Sutlej, and Beas rivers to India and the water from the Jhelum, Chenab and Sindh rivers to Pakistan. India's construction of a hydroelectric dam on the Chenab river, which flows from the Indian state of Jammu & Kashmir to Pakistan, has, according to Pakistani officials, reduced the water available for Pakistani agriculture.[61] This is of concern and an item of contention for Pakistan.

The Indus Water Treaty has allowed water sharing between India and Pakistan for many years. Environmental changes brought on by climate change may either threaten this cooperation or may enhance it. There is concern by some in Pakistan that India will use its upstream dams to manipulate water flow to its advantage and to Pakistan's disadvantage. It is thought that Himalayan glacier melt will increase flows of water in the nearer term and that this could lead to flooding in Pakistan if India does not use its upstream dams to mitigate this. It is also thought by some in Pakistan that after this period of thaw of Himalayan glaciers that there will be less water and that India may use its dams to hold back water from Pakistan.[62]

Tensions over the water resources may place more emphasis on Pakistani support of insurgents that are trying to break Kashmir away from India, which could potentially escalate to war given the history of conflict between India and Pakistan. Conversely, they could also lead to enhanced cooperation. Pakistani President Asif Ali Zardari wrote in 2009 that "The water crisis in Pakistan is directly linked to relations with India. Resolution could prevent an environmental catastrophe in South Asia, but failure to do so could fuel the fires of discontent that lead to extremism and terrorism."[63]

ISSUES FOR CONSIDERATION

Pakistan is of great interest to the national security of the United States. U.S. interests are primarily focused on Pakistan's ability to control its territory to prevent it from being used as a haven for anti-American terrorists and prevent inter-state conflict with India that would be regionally destabilizing or worse, given their nuclear weapons.[64] As a result, a stable Pakistan that can extend its full writ throughout the land in order to prevent various regions of Pakistan from becoming ungoverned areas from which anti-Western terrorists can operate is of direct national security importance. To this end, U.S. assistance may increasingly take into consideration the potential negative impact that environmental stress, particularly water shortages and its impact on food security, may have as a threat multiplier.

There is a potential need to prevent these exacerbating environmental factors from becoming the tipping point leading to a more chaotic situation in Pakistan. In April 2010 Special Representative for Afghanistan and Pakistan Richard Holbrooke stated that the United States was indeed putting more emphasis on energy and water issues in its assistance programs with Pakistan.[65] (For an in-

depth discussion of U.S. assistance to Pakistan see CRS Report RL33498, *Pakistan-U.S. Relations*, by K. Alan Kronstadt.) It has been argued by at least one source that because water shortages present the greatest future threat to the viability of Pakistan, U.S. development assistance would be best focused on the improvement of Pakistan's water infrastructure.[66]

End Notes

[1] National Security and the Threat of Climate Change, Center for Naval Analysis, http://securityand climate.cna.org/.

[2] Cuts in electricity have lead to several protests in Lahore and Karachi, among other regions.

[3] The Obama Administration sought nearly $1.6 billion specifically for Pakistan. Highlighted program areas to receive major funding increases include education ($364 million in FY2010); health care ($180 million); private sector competitiveness ($129 million); economic growth in the agriculture sector ($124 million); and rule of law and human rights ($39 million). See CRS Report RL33498, *Pakistan-U.S. Relations*, by K. Alan Kronstadt, for more information.

[4] J.R. McNeill, "Can History Help Us With Global Warming?" p. 35 in Kurt Campbell ed. *Climatic Cataclysm: The Foreign Policy and National Security Implications of Climate Change* (Washington: Brookings Institution Press, 2008).

[5] Thomas Homer-Dixon ed. *Ecoviolence: Links Among Environment, Population and Security* (Boulder: Rowman and Littlefield Publishers, 1998).

[6] For more detailed information on Pakistan water issues see Michael Kugelman and Robert Hathaway, eds., *Running on Empty: Pakistan's Water Crisis*, Woodrow Wilson Center, 2009.

[7] Thomas Homer-Dixon ed. *Ecoviolence: Links Among Environment, Population and Security* (Boulder: Rowman and Littlefield Publishers, 1998), p. 9.

[8] Jared Diamond, *Collapse: How Societies Choose to Fail or Succeed* (New York: Viking Press, 2005), p. 11.

[9] See M. Kugelman in the introduction to *Running on Empty: Pakistan's Water Crisis* (2009), published by the Woordrow Wilson International Center for Scholars. The paucity of water rights law in Pakistan means that landowners have better defined rights to water than do the

landless; similarly, large wealthy farmers often receive more irrigation water than small, poorer farmers. The report is available at http://www.wilsoncenter.org/topics/pubs/ASIA_090422_Running%20on%20Empty_web.pdf.

[10] Government of Pakistan, Federal Bureau of Statistics, *Pakistan Social and Living Standards Measurement Survey 2005-06*, available at http://www.statpak.gov.pk/depts/fbs/statistics/pslm2005_06/water_supply_sanitation.pdf. Thirty percent of households have no toilet facilities. Sanitation coverage varies largely between rural and urban areas, with only 3% of urban households not having toilets, and 44% of rural households.

[11] While there has been a degree of controversy over the exact number of years that it will take for Himalayan glaciers to melt, there is general agreement among scientists that they are melting at an alarming pace.

[12] "Himalayan Glacial Melting Still a Threat," *United Press International*, http://www.upi.com/science.

[13] "Global Climate Change: The Current and Future Consequences of Global Change," NASA, http://climate.NASA.gov/effects/, and Intergovernmental Panel on Climate Change (IPCC), "Moving Toward the Fifth Assessment Report, http://www.ipcc.ch/.

[14] U.N., "Pakistan, Country Profile," 2002, http://www.un.org/esa/agenda21/natlinfo/wssd/pakistan.pdf; and FAO, "Food and Agriculture Indicators," 2006, http://www.fao.org/es/ess/compendium_2006/pdf/PAK_ESS_E.pdf. GDP estimates may vary depending in part on the inclusion of cotton production and fisheries.

[15] Muhammad, D., "Pakistan, Country Pasture/Forage Resource Profile," 1998 (data last updated in October 2006), http://www.fao.org/ag/AGP/AGPC/doc/Counprof/Pakistan/Pakistan.htm.

[16] Comments from presenters at "Hunger Pains: Pakistan's Food Insecurity," Woodrow Wilson International Center conference, Washington, DC, June 3, 2009.

[17] U.N., "Pakistan, Country Profile," 2002, http://www.un.org/esa/agenda21/natlinfo/wssd/pakistan.pdf; and Library of Congress, "Country Profile: Pakistan," February 2005, http://lcweb2.loc.gov/frd/cs/profiles/Pakistan.pdf.

[18] Comments from presenters at "Hunger Pains: Pakistan's Food Insecurity," Woodrow Wilson International Center conference, Washington, DC, June 2009; U.N., "Pakistan, Country Profile," 2002, http://www.un.org/esa/agenda21/ natlinfo/wssd/pakistan.pdf.

[19] FAOSTAT, "Food and Agriculture Indicators," 2006, http://www.fao.org/es/ess/compendium_2006/pdf/ PAK_ESS_E.pdf. Data are reported for 2004.

[20] FAO, *FAO Statistical Yearbook*, 2004, http://www.fao.org/es/ess/yearbook/vol_1_2/pdf/Pakistan.pdf. Total land area is about 77.1 million hectares.

[21] FAO Terrastat and 2000 World Soil Resources Report, Table A3d, ftp://ftp.fao.org/ agl/agll/docs/wsr.pdf; and Iqbal, M.M., and A.M. Khan, "Climate Change Challenges faced by Agriculture in the Punjab," August 30, 2008. Based on the Aridity Index (AI), the U.N. Environment Programme's (UNEP) classification of climatic zones quantifies precipitation deficit over atmospheric water demand. AI equals the ratio between mean annual precipitation and mean annual potential evapotranspiration (or the amount of water that would be lost from water-saturated soil by plant transpiration and direct evaporation from the ground).

[22] Organic matter (OM) is an important measure of soil health. About 3%-4% OM is typical for cultivated, temperate-region soils.
[23] U.N., "Pakistan, Country Profile," 2002, http://www.un.org/esa/agenda21/natlinfo/wssd/pakistan.pdf; and Pakistan Mission to the United Nations, "Pakistan Country Profile," April 1997.
[24] Iqbal, M.M., and A.M. Khan, "Climate Change Challenges faced by Agriculture in the Punjab," Seminar presentation, August 30, 2008, http://cmsdata.iucn.org/downloads/pk_cc_mohsin.pdf.
[25] FAO, "Nutrition and Consumer Protection, Pakistan," http://www.fao.org/ag/AGN/nutrition/PAK_en.stm. Based on surveys of average daily requirements during 1984-1985 and 1987-1988. No recent nationwide food consumption surveys or information on the adequacy of present food consumption are available.
[26] FAOSTAT, "Food and Agriculture Indicators," 2006, http://www.fao.org/es/ess/compendium_2006/pdf/ PAK_ESS_E.pdf. Data are reported for 2004.
[27] FAOSTAT, "Food Security Statistics, Pakistan," http http://www.fao.org/faostat/foodsecurity/Cous_en.htm. Data are reported for 2001-2003 period.
[28] FAO, "Nutrition and Consumer Protection, Pakistan," http://www.fao.org/ag/AGN/nutrition/PAK_en.stm.
[29] Ibid; and FAOSTAT, "Food Security Statistics, Pakistan," http://www.fao.org/faostat/foodsecurity/Cous_en.htm. Data are reported for 2001-2003 period.
[30] FAO, "Nutrition and Consumer Protection, Pakistan," http://www.fao.org/ag/AGN/nutrition/PAK_en.stm.
[31] Comments from presenters at "Hunger Pains: Pakistan's Food Insecurity," Woodrow Wilson International Center conference, Washington, DC, June 3, 2009.
[32] It had a magnitude 7.6 on the Richter scale.
[33] U.S. Agency for International Development, South Asia Earthquake, Fact Sheet #44, at http://www.usaid.gov/ our_work/humanitarian_assistance/disaster_assistance/countries/south_asia/fy2006/southasia_eq_fs44_08-25-2006.pdf.
[34] "Pakistan's Current Monsoon Leads to an International Emergency Appeal for Assistance," http://www.examiner.com.
[35] "Floods Ravage Pakistan Valley That Endured Taliban," *Associated Press*, August 1, 2010.
[36] Glacial Lake Outburst Floods results from glacial dams bursting. Glacial dams form when glaciers melt and leave pools of water behind dams formed from silt and debris. Melting glaciers and deforestation are thought to be main causes for glacial lake formation.
[37] European Commission on Humanitarian Aid, *Managing Climate Risk: Glacial Outburst Floods*, United Nations Development Program, 2010, http://www.managingclimaterisk.org/index.php?menu_id=2&submenu_id=102& pagetype_menu=2&content_id=SMN-2.
[38] Ashfaq Yusufzai, "Militants Gain legitimacy Through Quake Relief," http://ipsnews.net.
[39] "Pakistan's Daunting—and Deteriorating—Demographic Challenge," January 18, 2010, http://www.populationmedia.org.
[40] "Pakistan at Extreme Food Security Risk," *Dawn*, September 8, 2009.
[41] Josette Sherman, "How to End Hunger," *The Washington Quarterly*, April 2010.

[42] Peiris, V., and K. Jones, "Pakistan Roiled by Flour and Electricity Shortages," http://www.buzzvines.com.

[43] Myra Imran, "Food Insecurity Haunts 45 Million in Pakistan," *The News,* October 16, 2009.

[44] "Urban Food Security Deteriorating," U.N. Office for Coordination of Humanitarian Affairs, IRIN, January 8, 2008.

[45] William Branigin, "U.S. to Provide Millions for Pakistani Refugees," *The Washington Post,* May 20, 2009.

[46] "Pakistan Politics: Power Shift," *Economist Intelligence Unit*, April 9, 2010.

[47] "Country Report: Pakistan," *Economist Intelligence Unit*, April 2010.

[48] "Pakistan Slips Five Places in Transparency International Corruption Index," *Pakistan Conflict Monitor*, November 18, 2009.

[49] Peter Gizewski and Thomas Homer-Dixon, "Environmental Scarcity and Violent Conflict: The Case of Pakistan," Occasional Paper, Project on Environment, Population and Security, April 1996.

[50] "Pakistan," *CIA World Factbook,* April 2010.

[51] Ximena Ortiz, "Food Follies," *National Interest*, April 24, 2009.

[52] For further information see CRS Report RS22983, *Pakistan's Capital Crisis: Implications for U.S. Policy*, by Michael F. Martin and K. Alan Kronstadt.

[53] "Pakistan: Country Report," *The Economist Intelligence Unit*, April, 2010.

[54] The direct importance of Pakistan to American security was highlighted most recently in early May 2010 as it became apparent that an immigrant from Pakistan, Faisal Shahzad, had received training in Waziristan before he attempted to detonate a car bomb in Times Square, New York, on May 1, 2010.

[55] "Militancy and Conflict Across the FATA and NWFP," *New America Foundation*, April 19, 2010. http://www.pakistanconflictmonitor.org

[56] "Government Desperate to Construct New Dams," *The Nation*, May 7, 2010.

[57] "Kalabagh Dam May Help Save Rs33b a Year," *Euclid Infotech*, May 7, 2010.

[58] "Indus River Delta-Arabian Sea Mangroves," http://www.worldwildlife.org.

[59] Altaf Memon, "Evaluation of Impacts on the Lower Indus River Basin Due to Upstream Water Storage and Diversion," University of Maryland July 2004.

[60] Altaf Memon, "Evaluation of Impacts on the Lower Indus River Basin Due to Upstream Water Storage and Diversion," University of Maryland July 2004.

[61] "India and Pakistan: Troubled Water," ISN Zurich, http://www.isn.ethz.ch.

[62] Ali Khizar Aslam, "Politics of water Insecurity," *Business Recorder,* April 21, 2010.

[63] Asif Ali Zardari, "Partnering with Pakistan," *The Washington Post*, January 28, 2009.

[64] For additional information see CRS Report RL34763, *Islamist Militancy in the Pakistan-Afghanistan Border Region and U.S. Policy*, by K. Alan Kronstadt and Kenneth Katzman.

[65] "Remarks by Richard Holbrooke, *Special Representative for Afghanistan and Pakistan, Federal News Service*, April 20, 2010.

[66] Anatol Lieven with John Hulsman, "Developmental Realism," *New America Foundation,* January 17, 2007.

In: Environmental Stress in Pakistan … ISBN: 978-1-61470-224-5
Editor: Omar Siddiqui

Chapter 2

FLOODING IN PAKISTAN: OVERVIEW AND ISSUES FOR CONGRESS*

K. Alan Kronstadt, Pervaze A. Sheikh and Bruce Vaughn

SUMMARY

Pakistan experienced a catastrophic natural disaster that has precipitated a humanitarian crisis of major proportions. Widespread flooding affected about 20 million Pakistanis and inundated an area the size of Florida within the country. Congressional interest in the flooding stems from the significant humanitarian and economic implications for the country, and the security implications for U.S. interests in the region. The World Bank and Asian Development Bank have estimated that the flooding has caused $9.7 billion in damages. While this figure might still be preliminary, it is almost certain that the negative effects of this crisis will be felt for many years to come.

The floods stemmed from abnormally heavy rains during the monsoon season in July and August, 2010. This led to flooding in the Indus River Basin which traverses Pakistan from north to south. Excess

* This is an edited, reformatted and augmented version of a Congressional Research Service publication, CRS Report for Congress R41424, from www.crs.gov, dated November 18, 2010.

water led the Indus River and its tributaries to breach their levees and inundate adjacent and downstream floodplains. Approximately 2000 people were believed to have been killed by the flooding. One fifth of the country was submerged, and an estimated eight million Pakistanis were displaced from their homes. The number of people affected were significantly greater than several major disasters around the world since 2000. Little clean drinking water was available for many of the people who were affected and remains a problem today. Many of those affected, particularly children, face potential disease outbreaks, particularly diarrhea and cholera. The catastrophic loss of livestock and crop lands and extensive damage to the country's infrastructure are projected to have long-term negative effects on Pakistan's food security and economic performance.

Pakistan is at the center of several crucial U.S. interests, including fighting terrorism and religious militancy, seeking stability in neighboring Afghanistan, and promoting nuclear non-proliferation, among others. The aftermath of the floods can affect broad political and strategic dynamics in Pakistan and the region in a number of ways. The crisis may undermine the already waning legitimacy of the civilian government by demonstrating its ineffectiveness to large numbers of Pakistanis in need of public services, while improving the status of Pakistan's powerful military by the more visible role it played in providing disaster relief. It may also provide militants an opportunity to garner favor with affected communities by giving militants an opportunity to demonstrate that they can provide assistance in areas where the government is absent. The crisis has also diverted attention and resources from other national priorities, at a time when Pakistan remains financially strapped.

U.S. interests are served by a stable Pakistan that can effectively rule all its territory. Any crisis on a scale of the present floods that undermines the Pakistani state's ability to control its territory has the potential to undermine U.S. interests. The inability of Pakistan to fully extend its authority into areas along its northwest frontier with Afghanistan has allowed Islamist militants hostile to the United States to find refuge. The flooding diverted Pakistani resources and focus away from its struggle with Islamist militants. This has the potential to indirectly affect U.S. military involvement in Afghanistan by taking pressure off militants on the Pakistani side of the international frontier. On a positive note, the crisis presents the United States with an opportunity to improve its poor image among Pakistanis through provision of humanitarian assistance. Congress will play an important role in overseeing such assistance in the near term, and broad foreign assistance strategies for rebuilding infrastructure and other development goals in the medium and long run.

INTRODUCTION

Widespread flooding in Pakistan affected millions of people and caused a significant humanitarian disaster with widespread economic and security consequences for the country. A stable, democratic, prosperous Pakistan actively combating religious militancy is considered vital to U.S. interests. The humanitarian effects of the flooding along with its security consequences generated considerable concern in the United States and many countries around the world. The flooding in Pakistan, which started in late July 2010, reportedly affected over 20 million people. Over 1.9 million households were destroyed or damaged.[1] Approximately 2,000 people died and over 3,000 were injured.[2]

In terms of affected persons, the humanitarian impact of the 2010 floods in Pakistan is estimated to be larger than other recent natural disasters experienced globally. The estimated 20 million people affected by flooding in Pakistan is more than the estimated number of people affected by the 2010 earthquake in Haiti, the 2005 tsunami in the Indian Ocean littoral region, and the 2005 South Asia earthquake combined.[3] The death toll increased after floodwaters receded in some areas, although specific number of deaths directly attributable to flooding is difficult to ascertain. Lost shelter and food staples, limited access to clean water, and unsanitary conditions from standing water and submerged lands are immediate humanitarian issues. Cases of malnutrition, gastroenteritis, diarrhea, skin infections, cholera, typhoid, malaria, and hepatitis have been reported. The Pakistan government established a National Oversight Disaster Management Council to improve the transparency in the distribution of flood relief. However, humanitarian relief efforts were tempered by initially low levels of donations, difficulties in accessing victims, the large geographical extent of the disaster, and security concerns in some regions.[4]

Some medium and long-term consequences of flooding have been projected. Lost livelihoods for farmers (e.g., not being able to plant next season's crop) and a diminished food supply is a key concern. The flooding destroyed crops, food stockpiles, livestock, seeds, structures, and equipment. Food prices dramatically increased since the flooding, putting an economic strain on the entire population. Rebuilding damaged and destroyed housing is also expected to affect the return of displaced populations. As tensions mount and the struggle to stabilize persists, public confidence in government efforts to provide assistance will be tested. Long-term consequences such as damaged infrastructure (e.g., bridges, roads, electrical plants, and schools) are also expected to hamper recovery efforts and have a lasting effect on the country.

Before the flooding, top-tier U.S. concerns regarding Pakistan included regional and global terrorism; stability in neighboring Afghanistan; domestic political stability and democratization; nuclear weapons proliferation and security; human rights protection; and economic development. The 2010 floods have exacerbated these concerns. The effects of flooding on the stability of the Pakistani government and its ability to suppress and contain militants within their borders is an issue of concern for the United States and others.[5]

Congressional Interest

Congressional interest regarding the 2010 flooding in Pakistan stems from multiple factors, including the U.S. response to the humanitarian needs of flood victims, potential cost of reconstruction after the flood subsides, the composition of aid given by the United States to Pakistan to garner its support for U.S. operations in Afghanistan, the disabling of Taliban inside Pakistan's borders, and the security concerns arising from a weakened Pakistan with a potentially disenfranchised and dissatisfied population.

Public opinion surveys in Pakistan before the floods present two issues of concern for the United States: (1) that the percentage of Pakistanis polled who are worried about extremist groups taking control of Pakistan has declined from 69% in 2009 to 51% in 2010; and (2) that about 68% polled view the United States unfavorably.[6] These numbers are expected to change after the floods as perception of a weak government grows and pressure to contain militants softens (see later sections of the report for more details). The United States, however, has an opportunity to improve its perception among Pakistanis. While humanitarian benefits are the priority, emergency aid and reconstruction efforts, if successfully implemented, may provide long-term diplomatic benefits for the United States in Pakistan.

Potential Actions for Congress

One potential issue for Congress and the Obama Administration is deciding what additional actions and funds are available, warranted, and effective. Many have suggested potential actions to address flooding in Pakistan. Some of those actions include

- Additional U.S. emergency relief funding, equipment, and supplies
- U.S. efforts to promote contributions and cooperation from other nations for emergency response and reconstruction
- Actions intended to support the Pakistan economy and trade[7]
- Debt relief
- Efforts to promote individual and corporate aid and contributions
- Measures to promote foreign-direct investment and public-private partnerships

CONTEXT OF THE 2010 PAKISTAN FLOOD

Heavy rains associated with monsoons began around July 22, 2010, and led to flash floods in the northwest and east of Pakistan. (See **Figure 1**.) Downpours continued for the rest of the month and persisted into August. The runoff caused the Indus River and its tributaries to breach levees and overflow into floodplains housing both rural and urban populations. At times river flows reached 40 times their normal levels. As floodwaters moved downstream, new areas of flooding emerged in southern provinces, such as Sindh. Intentional levee breaks were made in attempts to protect urban centers (e.g., Hyderabad); nonetheless, levee breaches affecting populations clusters (e.g., Thatta city) were still occurring in early September.

Table 1. Provinces in Pakistan Affected by Flooding (As of September 6, 2010)

Province	Deaths	Injured	Houses Damaged	Population Affected
Baluchistan	48	102	75,261	672,171
Khyber Pakhtunkhwa	1,156	1,193	200,799	4,365,909
Punjab	110	350	500,000	8,200,000
Sindh	199	909	1,058,862	6,988,491
AJ&K	71	87	7,108	245,000
Gilgit Baltistan	183	60	2,830	81,605
Total	1,767	2,701	1,844,860	20,553,176

Source: United Nations Office for the Coordination of Humanitarian Affairs, September 14, 2010.

The effects of the flooding covered a large geographical area in Pakistan and affected several provinces. (See Table 1 for September estimates.) The timing of the flooding effects varied for provinces. For example, in many affected regions in the north (e.g., Punjab and Khyber Pakhtunhwa provinces), affected populations have largely returned to their damaged communities and started rebuilding. However, in the south, such as in the Sindh province, stagnant floodwaters are preventing recovery, and significant displaced populations exist as of November 2010. It is reported that nearly 7.0 million people are affected by floods in Sindh province, with thousands trapped by flood waters and in need of assistance.[8] Areas inundated weeks after the onset of flooding suffered significant agricultural and property losses, and concerns about food, shelter, safe water, sanitation, health, and livelihood persist.

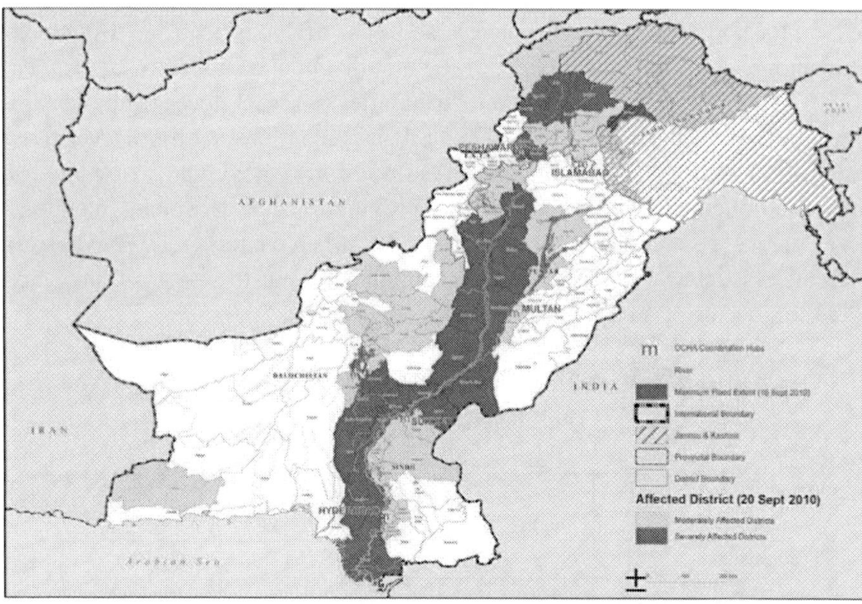

Source: United Nations Office for the Coordination of Humanitarian Affairs.

Notes: The designations employed and the presentation of material on this map do not imply the expression of any opinion whatsoever on the part of the Secretariat of the United Nations concerning the legal status of any country, territory, city, or area or of its authorities, or concerning the delimitation of its frontiers or boundaries. Dotted line represents approximately the Line of Control in Jammu and Kashmir agreed upon by India and Pakistan. The final status of Jammu and Kashmir has not yet been agreed upon by the parties.

Figure 1. 2010 Extent of Flooding in Pakistan, September 20, 2010.

GOVERNMENT OF PAKISTAN RESPONSE

The response of the Government of Pakistan to the flooding is being organized at the federal, provincial and district levels. The Economic Affairs Division (EAD) is coordinating donors and the Pakistan National Disaster Management Agency (NDMA) is responsible for overall coordination of disaster response efforts by both the government and the international community. The NDMA works closely with federal ministries, government departments, the armed forces, UN agencies and donors to mobilize, receive and deploy relief goods.[9] Relief activities are also being coordinated by provincial-level and district-level governments.

The primary Pakistani response to the flooding was initially and has continued to be from the military. The military moved its forces and equipment into the flooded regions to evacuate people and distribute supplies. The civilian government response was reportedly weak, due in part to the damage and destruction of local governmental assets and properties during the flooding. Some also blame coordination issues for a slow state response, such as the absence of civil-military coordination and ineffective civilian control over military-led efforts.[10] There were several reports of widespread criticism by the Pakistani people of the civilian government response to aid flood victims. Some, however, have defended the agency by stating that it was not structured to cope with a disaster of this magnitude.[11]

To help families rebuild, the Pakistani government is distributing debit cards to affected families. The Pakistan National Database and Registration Authority has distributed debit cards to over 900,000 families as of September 2010.[12] They aim to provide $1,150 dollars to each head of household in three separate disbursements to help families recover from the floods. Steps are being taken to include female-head of households in the process for those females who lost their husbands.

UNITED STATES RESPONSE

The United States is the largest donor of funding for relief efforts related to the flooding in Pakistan. Some funds are being converted from a portion of the civilian economic development assistance authorized by the Enhanced Partnership with Pakistan Act of 2009 (EPPA 2009, P.L. 111-73), which is appropriated primarily within the FY2010 base and supplemental appropriations (P.L. 111-117 and P.L. 111-212 respectively). To date, the United States is providing a total of $561.9 million for Pakistan relief and recovery, largely coming from International Disaster Assistance (IDA) and Food For Peace (FFP) funds. (See Table 2.) In-kind civilian and military support, such as the pre-fabricated steel bridges, halal meals, and air transport, amounts to another $89.1 million, according to USAID.[13]

Table 2. FY2010 Humanitarian Aid for Pakistan Floods (in millions of U.S. dollars)

FY2010 flood aid		FY2011 flood aid
USAID Office of Foreign Disaster	$115.0	$76.2
Assistance (OFDA)	(All IDA)	(all IDA)
USDA/Food for Peace (FFP)	$140.7	$90.0
	(IDA and title II)	
USAID Pakistan Assistance	$10.0	
	(all ESF funds,$2 million from OTI after transferred from ESF)	
USAID Pakistan Early Recovery	$69.8	$11.0
Assistance	(all ESF funds from FY2010 supp)	(all ESF)
State Dept. Bureau of Population,	$49.2	
Refugees & Migration (PRM)	(from ERMA & MRA FY2010 base)	
DoD	Not available	
Total	$384.7	$177.2

Source: Communications with the Office of the Director of Foreign Assistance, Department of State, and USAID, September 21, 2010.

Note: In addition, the U.S. has provided civilian and military in-kind aid valued at about $89.1 million.

Prior to the flooding emergency, the United States provided humanitarian aid to Pakistan to address the January 2010 landslides, conflict-affected displaced people (referred to as complex emergency funds), and pre-2010 earthquake expenses. (See Table 3.) A potential issue for Congress is whether the purpose of these funds, as well as the authorized economic development funds from the EPPA 2009, should be changed to reflect flood relief efforts and changed circumstances caused by the flood. If the funds are redirected, then there could be concerns as to how the original objectives of the funding will be met.

The United States Agency for International Development (USAID) is leading all U.S. government humanitarian response efforts to Pakistan. Nearly 30 U.S. military helicopters and four C-130 planes were involved in delivering humanitarian supplies. These included emergency food supplies, plastic sheeting, inflatable rescue boats, concrete-cutting saw kits, shelters, blankets, water storage bladders, and water filtration units among other things. Twenty-six U.S. mobile medical teams were providing health care services, which were needed because about 450 health facilities were destroyed or damaged by the floods, according to OCHA.

U.S. Private Donations

On August 18, 2010, Secretary Clinton announced that the Department of State had created a Pakistan Relief Fund for individuals, corporations, and others to contribute toward relief and recovery efforts in Pakistan. Donations can be made through either the Department of State or USAID websites. As of early September 2010, private sector donations amounted to $8.39 million, according to USAID. Individuals, corporations, and other organizations can send donations to the people of Pakistan by contributing to this fund at http://www.state.gov or http://www.usaid.gov/pakistanflooding/. In the United States, individuals also can send funds through mobile phones by texting "FLOOD" to 27722.

Table 3. FY2010-FY2011 U.S. Humanitarian Aid to Pakistan (in thousands of dollars)

Source	FY2009 Base	FY2009 Spring Supp	FY2010 Base	FY2010 Supp	FY2011 Base (CR)	Total
Appropriation Laws	P.L. 111-8	P.L. 111-32	P.L. 111-117	P.L. 111-212	P.L. 111-	
					242	
Int'l Disaster Asst. (IDA)						
Complex Emergency	$256.0	$16,569.2	$225.3	$1,500.0		$18,550.5
Earthquake		$193.5				$193.5
Landslide	$50.0					$50.0
Floods		$627.4	$22.6	$114,355.7	$76,181.2	$191,186.9
Total IDA	$306.0	$17,390.1	$247.9	$115,855.7	$76,181.2	$209,980.9
Food For Peace						
Complex Emergency transferred from IDA			$20,000.0			$20,000.0
Complex Emergency – Title II	$55,386.7		$70,820.8			$126,207.5
Floods – transferred from IDA			$120,329.8			$116,978.7
Floods – Title II			$20,399.1			$20,399.1
Floods Total			$140,728.9		$90,015.6	$230,744.5
Total Food For Peace	$55,386.7		$231,549.7		$90,015.6	$321,565.3

Source	FY2009 Base	FY2009 Spring Supp	FY2010 Base	FY2010 Supp	FY2011 Base (CR)	Total
Economic Support Fund (ESF)						
USAID Pakistan Flood Aid			$10,000.0[a]			$10,000.0
USAID Pakistan Flood early recovery aid				$69,805.0	$11,000.0	$80,805.0
Total ESF – Flood Aid			$10,000.0[b]	$569,805.0	$11,000.0	$90,805.0
State Dept – Population, Refugees & Migration						
Complex Emergency (ERMA)	$13,300.0					$13,300.0
Complex Emergency (MRA)	$1,300.0	$45,000.0[c]	$32,000.0			$78,300.0
Floods (ERMA)			$33,021.0[d]			$33,021.0
Floods (MRA)			$16,229.0			$16,229.0
Floods PRM Total			$49,250.0			$49,250.0

Source: USAID's Office of Foreign Disaster Assistance, Office of Food For Peace; Department of State's Bureau of Population, Refugees, and Migration, and the Foreign Assistance (F) Bureau.

Notes: IDA = International Disaster Assistance; ERMA=U.S. Emergency Refugee and Migration Assistance; MRA = Migration and Refugee Assistance.

[a] Of this, $2.0 million was transferred to Office of Transition Initiatives (OTI) for flood assistance.

[b] Of this, $10.0 million was appropriated within the Bridge Supplemental for FY2009 (P.L. 110-252).

[c] Includes $21,000 recovery funding from prior ERMA appropriations.

[d] Includes $819,000 recovery of prior year MRA funding.

International Response

Other countries are providing support for Pakistan flood relief efforts. China claims it was the first to contribute aid to the flood victims and has provided about $47.0 million.[14] The European Union has provided a total of about $450.9 million of cash and in-kind aid, including about $210.4 million from the European Commission Humanitarian Aid Department and sums from more than 20 countries. For example, Germany about $47.0 million cash and in-kind, Denmark about $24.4 million, Sweden about $25.5 million, Norway about $66.0 million, the United Kingdom about $209.0 million, and France about $4.3 million.[15] As of September 8, 2010, Japan has provided a total of $25.6 million in cash and relief commodities.[16] As of the end of August 2010, Australia has provided $75 million and Canada has pledged about $52 million U.S. dollars with an additional $1.4 million from a Canadian humanitarian coalition.[17] India also has provided $5 million of aid and is offering more, although receiving aid from India is controversial in Pakistan.

Overview of the Humanitarian Response

The humanitarian implications of the flooding in Pakistan are vast. Humanitarian issues such as health, sufficient food, water, and shelter, and displaced populations exist. Various offices under the United Nations are coordinating with the Government of Pakistan to address humanitarian needs in the country. The United Nations released The Pakistan Initial Floods Emergency Response Plan (PIFERP) on August 11, 2010, seeking approximately $460 million to respond to the immediate relief needs of flood-affected people. The money is to be used to enable international partners (U.N. and non-governmental organizations) help the Pakistani government address residual relief needs and recovery needs for the next year. This plan covered 12 humanitarian clusters including water and sanitation, health, shelter, agriculture, food, community restoration, protection, education, nutrition, logistics, coordination, and camp management. Funding needs for each cluster are identified in the plan, and planned activities under each cluster are listed. Coordinating entities for each cluster are listed and range from the Food and Agriculture Organization (FAO) for agricultural needs to the United Nations Development Program (UNDP) for community restoration needs. Monitoring and methods for evaluating progress are also included in the plan. In November 2010, a revised plan was issued by the U.N.[18] This plan seeks $1.9

billion for the above-mention sectors to be disbursed from August 2010 to August 2011.[19] The following sections discuss the U.S. and Pakistani government response to the flooding.

Food Security

Prior to the recent flooding, poverty and hunger in Pakistan were widespread and especially prevalent in rural areas. Nearly two-thirds of the population and 80% of the country's poor (about 35 million people) live in rural parts of the country. Women in Pakistan, who have the lowest socio-economic status in South Asia, are often among the most vulnerable populations, along with children. The recent global food price and economic crises of 2008-2009 exacerbated poverty and food security issues in Pakistan. FAO estimated that an additional 17 million people became food-insecure as a result of food price inflation in Pakistan over the past few years, and that the poorest households are now spending more than 70% of their incomes on food. Even before the flooding, FAO had estimated that about 60 million people were food-insecure in Pakistan, which accounts for about half of the country's population. While the full extent of damage from the summer 2010 flooding has not yet been fully quantified, the direct and future losses are likely to impact national production of staple crops, such as wheat and rice, and affect the food security of millions of people.

According to the PIFERP, distributing food assistance to flood victims is among the top priorities for the overall immediate flood response. The World Food Programme (WFP) is the lead U.N. agency responsible for coordinating and leading the efforts to address the immediate food needs of flood-affected populations in the hardest-hit areas of the country. The key expected outcomes from the food relief efforts will be to:

- Stabilize and/or improve food consumption over the assistance period for six million flood-affected people per month during the next 12 months,
- Increase the supply of nutritious food to infants and young children, and
- Prevent additional losses of surviving livestock.

WFP is working with over 30 NGOs to provide food to between 125,000 and 150,000 people per day in order to meet their ultimate goal of reaching more than six million people with food aid over the next 12 months. Many communities are inaccessible by land because of remaining high floodwaters and/or because roads and bridges have been washed away. WFP is transporting food by mule and air transport (primarily helicopter) to inaccessible populations.

WFP is requesting approximately $600 million to feed people from August 2010 to July 2011, of which a portion will be spent on maintaining surviving livestock. Since November 2010, the WFP has received $255 million, of which $167 million is from the United States.

Effects on Health

Immediate and long-term health risks caused by the flood are high. Flooding can increase the spread of water-borne diseases when access to clean drinking water is compromised. Likewise, standing water caused by flooding can serve as breeding grounds for mosquitoes, increasing the potential for vector-borne diseases. These risks are heightened when there is significant population displacement, abnormal overcrowding, and a reduction in disease control activities.[20] Children are particularly vulnerable in these circumstances.

Access to health services and medicines is of key concern in Pakistan's flood-affected areas, as is the restoration of the country's public health infrastructure. The World Health Organization (WHO) Pakistan Health Cluster reports that 514 health facilities in flood-affected areas have been damaged or destroyed, including several hospitals.[21] Of key concern is the insufficient access that women have to reproductive health services and the limited number of female health workers available.[22] Poor hygiene and sanitation conditions in flood affected areas represent a major risk factor, particularly in settlements where individuals have informally or officially gathered. As of late August, UNICEF reports indicate that approximately 3.5 million flood survivors only have access to contaminated water.[23]

Through the end of September 2010, over 6 million individuals received flood-related medical consultations. The main conditions reported are acute diarrhea, acute respiratory infections, skin diseases, and suspected cases of malaria.[24] Cholera is endemic in Pakistan and the threat of an outbreak in flood-affected areas remains critical; as of October 2010, a total of 99 confirmed cases were reported.[25] The WHO Health Cluster is currently treating all cases of acute watery diarrhea and severe dehydration as if they were cholera. The present health issues are both preventable and treatable;

however, if clean water is not supplied and environmental hazards not addressed, WHO projects that a worst-case scenario includes 1.5 million cases of diarrheal diseases (including up to 140,000 cases of cholera) and up to 100,000 cases of malaria over the next three months.[26] In addition, there is increasing concern over dengue fever, Crimean-Congo haemorrhagic fever (CCHF), measles and polio, and malnutrition among children and pregnant women.[27] Displaced people returning home to areas with fewer health care facilitates than in the settlements now represent a new high-risk health group.[28] Mental health care for survivors dealing with trauma is also an acknowledged priority.

Displaced Populations

A major concern is the fate of displaced populations in Pakistan. Several regional populations of Pakistanis have undergone cycles of displacement from militancy and military operations. Before the flooding, for example, the Pakistani military initiated the return of approximately 2.9 million displaced people in the Malakand region of the Khyber Pakhtunkhwa (KPK).[29] With the flooding, many of the returning displaced populations have been displaced yet again with additional damage to infrastructure and households. In addition, nearly 90% of the 1.7 million registered Afghan refugees reside in flood-affected areas, many could be displaced and their resettlement is uncertain. The government is attempting to register internally displaced people, but several obstacles are preventing a comprehensive tally.

The recent flooding has damaged over 1.9 million houses, and pushed a similar amount of people to schools, colleges, and other makeshift sites for accommodations. According to the United Nations Office for the Coordination of Humanitarian Affairs (OCHA), a large need for emergency shelter exists. The majority of adults in these sites are women. Many of the men have endured the flooding and gone back to their property to salvage belongings and deter squatters.

There have been some reports about the potential of displaced people to cause political unrest in some cities. For example, thousands of people displaced by the floods, most of them ethnic Sindhis, are arriving in Karachi. According to some, this could be adding a volatile new element to the political dynamic there. While Sindhi nationalists are welcoming them, opponents such as Muttahida Qaumi Movement (M.Q.M.) members warn that they will create more violence.[30]

Comparing Responses to Haiti and Pakistan: Lessons Learned

Some advocate that lessons learned from the earthquake disaster in Haiti could be applied to assisting Pakistan. However, differences in the area, people affected, and magnitude of the disaster might make comparisons tenuous. Haiti and Pakistan experienced megadisasters in 2010, each with different magnitudes and areas affected. For example, the floods in Pakistan affected more people (over 20 million) than the earthquake in Haiti (approximately 3 million). However, more people were killed and instantaneously displaced in Haiti than in Pakistan. Earthquakes are instantaneous and cause large numbers of fatalities compared to floods. Floods generally happen over time and cause less fatalities. Relief operations were implemented in both Pakistan and Haiti following natural disasters; however, the effects and dynamics of the disasters, areas affected, and timing of the disasters, in part, made their relief responses different. Many contend that aid donations from around the world were slower to materialize for Pakistan than other natural disasters, such as the earthquake in Haiti. For example, twice as much aid has been raised for Haiti (approximately $3.3 billion) than for Pakistan (approximately $1.6 billion). Indeed, after two weeks of the appeal for aid for Haiti and Pakistan, approximately $157 dollars was raised per affected Haitian compared to $15 dollars per affected Pakistani.[31]

Some potential reasons for the difference in donor response between these two disasters include

- The Haiti earthquake was an emergency situation that happened overnight and with immediate and visible tragedy, while the flooding in Pakistan occurred more slowly, over a longer period of time.
- There was less U.S. media coverage of Pakistan as compared with coverage in Haiti. Perhaps this was due to the suddenness of the earthquake in Haiti, and the ability of the media to access parts of Pakistan initially affected by the flooding.
- The high number of deaths and injuries associated the earthquake in Haiti (approximately 220,000 deaths and 300,000 injuries) compared with Pakistan (approximately 2000 deaths and 3,000 injuries) might have made the necessity to donate greater for Haiti than Pakistan.
- The association of Pakistan with Islamic extremists and the war on terror might have limited appeal. Further, some might have had concerns over government corruption and the oversight of donations.

- The difference in U.S. donations might have been related to the proximity of the two countries to the United States. Haiti is closer to the United States and has a large expatriate population in the United States.
- Donor fatigue from the earthquake might have limited donations and resources for Pakistan.

Lessons Learned from Haiti

Response to the earthquake disaster in Haiti might provide some lessons learned for the flooding in Pakistan despite the differences in the disasters. This section discusses lessons learned in Haiti and provides some examples of their potential application to Pakistan.

The delivery of aid to Haiti after the earthquake was criticized by some as being too slow. Others countered this notion by arguing that the weakened capacity of the Haitian government, critically damaged infrastructure, and logistical challenges posed by the influx of massive aid into a city largely destroyed by the earthquake all contributed to delay and difficulties on the ground. In its six-month review, the U.N. Inter-Agency Standing Committee (IASC) refers to the "perception of a coordination deficit in the initial phase of the response operation" and a sense of a lack of leadership, but also points to significant improvements that were made over subsequent months.[32]

The United States and other donors continue to work closely with the government of Haiti, the United Nations, NGOs, and the private sector through the U.N. cluster system.[33] Within days of the earthquake, relief activities were organized into twelve clusters led by various agencies, including emergency shelter, food, and health. Coordination through the cluster system, the convening of the Humanitarian Country Team for overall strategic guidance, the deployment of more senior humanitarian staff, and the establishment of other coordination structures, such as those that brought together military and humanitarian actors, reportedly eased decision-making and resource-tasking in the months following the earthquake.

Potential lessons learned by the international community that could be applied to Pakistan include[34]

- Engaging more with civil society, local authorities, and displaced populations for greater understanding of the operating context and sustainable delivery of assistance;
- Understanding and working with various actors outside the immediate humanitarian context, not only the military, but also the private sector;

- Engaging better with the government of the affected country and civil society partners;
- Providing cash for relief and buying locally wherever possible;
- Creating transitional shelter that can be turned into permanent dwellings, rather than focusing on emergency shelter and neglecting permanent shelter;
- Starting recovery as soon as possible without negatively affecting the relief effort; and
- Preparing for land-ownership disputes and not expecting the disaster response to resolve underlying political problems such as corruption, poor governance, underdevelopment, and social inequality.

Evaluations of the relief response in Haiti will likely continue to be conducted and debated as the humanitarian and recovery efforts move ahead. Some experts remain concerned about bureaucratic red tape in the humanitarian response, the capacity of the Haitian government, the role of the United States, and overall coordination issues between and among members of the international community, including the United Nations. These concerns can also be applied to Pakistan, with the added challenge of responding to a disaster that covers a large area with little viable infrastructure for transportation. Responses to earthquakes and flooding of the magnitude experienced in Haiti and Pakistan is almost certain to run into many obstacles because the challenges on the ground are so daunting. While managing expectations of what is possible under these circumstances is important, so too are the observations and lessons learned, which with time and hindsight may benefit the actions and plans of ongoing relief efforts in Haiti, Pakistan, and elsewhere.

Implications of Flooding on Selected Sectors of Pakistan

The floods are expected to have a long-term negative effect on the development prospects for Pakistan. There is considerable damage to infrastructure and agriculture, among other sectors. The United Nations, World Bank, and Asian Development Bank are conducting a needs assessment for Pakistan. Initial reports state that the need for recovery and reconstruction could reach $9.7 billion. This section discusses some potential implications of the flooding on selected sectors in Pakistan including energy and infrastructure, the economy, and security.

Implications on Energy and Infrastructure

Rebuilding challenges include reconstruction of destroyed or damaged housing and infrastructure, including electrical generation and distribution, roads, bridges, rail lines, levees, dams/barrages, and irrigation works. Damage estimates include more than 5,000 miles of primary and secondary roads, 400 bridges, 400 miles of railways, 11,000 schools, and 200 health facilities, with damage being particularly severe in northern regions like the Swat Valley.[35] The difficulty, time, and expense of rebuilding this lost infrastructure is likely to be substantial. For example, the World Bank and Asian Development Bank jointly estimate that the floods caused $9.7 billion of damages.[36] That figure includes damage across several sectors, including loss of physical infrastructure and other economic losses. The full study has not yet been released, but researchers have commented that agriculture and roads were particularly hard hit.[37] Two U.S. university researchers using primarily data on the flood characteristics and pre-flood economic data estimated damages between $5.1 billion and $7.1 billion to building and transportation infrastructure and $2.12 billion in losses from the disruption of trade.[38]

Floods shut down some electricity, oil, and gas facilities. According to various media reports, floods closed approximately 3 gigawatts of power generation capacity.[39] (Pakistan's maximum power generation capacity before the flood was around 19 gigawatts.[40]) Flooding damaged generation facilities and transmission infrastructure, and it cut off power plants from their supply of generation fuels such as oil and natural gas.[41] Output at refining and natural gas facilities also was curtailed due to transportation disruptions.[42] Most of this energy production and transportation capacity has been restored.[43, 44] However, the damage has highlighted and exacerbated Pakistan's pre-existing energy problems. Prior to the floods, the country was already suffering from a shortage of electricity generation capacity and rolling blackouts.[45] The cost of recovering from flood damage sets back efforts to improve electricity supply. Prior to the floods, the Obama Administration had pledged $185 million toward various measures to improve Pakistan's energy supply.[46]Assessments of the value and replacement cost of other infrastructure (e.g., energy, medical, and educational facilities) or a reconstruction timeline were not available as of late October. Among the challenges for rebuilding the infrastructure are prioritizing reconstruction actions; the availability of materials, equipment, and expertise; and mechanisms for oversight of construction and use of funds.

Implications for Agriculture

Agriculture is one of the primary mainstays of Pakistan's economy. It accounts for approximately 23% of GDP, employs about 43% of the labor force and provides about 60% of the country's export earnings.[47] Arable crops, livestock, and fishing and forestry represent 65%, 31%, and 4% of Pakistan's agricultural GDP, respectively. Pakistan typically has two major growing seasons, *Rabi* (winter crop, spring harvest) and *Kharif* (summer crop, fall harvest).[48] The *Kharif* crop is also called the summer or monsoon crop because it is grown during the time of the southwest monsoons, which typically occur from July to October. During the *Kharif* season, agricultural activities take place in rain-fed and irrigated areas. During the *Rabi* season, agricultural activities take place only in the irrigated areas. Nearly 80% of the cropped area is irrigated, producing about 90% of Pakistan's total farm output.

Pakistan's agriculture sector has suffered severe losses as a result of the torrential rain and flooding of July and August 2010. The majority of adverse flooding impacts have occurred in four provinces: the Punjab, often called the breadbasket of Pakistan, and Khyber Pakhtunkhwa (KPk) in the north, and the provinces of Sindh and Baluchistan in the south, all of which have significant crop production. Approximately 80% of people in the flood-affected areas depend on agriculture for their livelihood. The affected populations have suffered severe crop, livestock, and grain stock losses, though assessments of medium and longer term impacts on the agricultural sector in Pakistan are still ongoing.

Crops

Damage assessments by U.N. agencies led by the FAO estimate that about 3.3 million hectares countrywide of standing crops, including rice, maize, cotton, sugarcane, fruit orchards and vegetables, have been damaged or lost completely due to flooding, with about 1.3 million hectares affected in the four hardest hit provinces.[49] This represents about 14% of the total cropped area in 2008.[50] The cereal crops of the current *Kharif* or monsoon season (including rice, maize, sorghum, and millet), planted in May/July and scheduled to be harvested from September onward, typically account for approximately 35% of the annual national cereal production. Reports from FAO and the Asian

Development Bank have estimated that as much as 25% of the cotton crop has been affected. Pakistan consequently may be a net importer rather than an exporter of cotton in 2010 in order to support its textile industry. Substantial losses of important export crops such as cotton, sugarcane, and rice, will have a large negative impact on the country's trade balance and household incomes, while losses in the production of staple cereals will exacerbate household food security issues in the country.

While the floods are causing severe negative effects on agricultural production in the current season, the damage and impacts will likely have broader implications for future agricultural production and food security in Pakistan. Wheat is an example of this issue.

Wheat is the main staple in Pakistan, providing about 35% of the average per capita calorie requirement in 2008. Wheat is mainly irrigated and contributes approximately two-thirds of the annual national cereal production. It is cultivated primarily during the *Rabi* season, when it is typically planted in October/November and harvested in April/May. Official final estimates of wheat production from the 2009/2010 *Rabi* season, which was successfully harvested prior to the flood, were close to 23.9 million tonnes, a near record amount. However, the flood waters have resulted in serious losses of wheat stocks that are held at the household level. These reserves serve as the primary source of food for most rural families until the next harvest. Estimates made by FAO indicate that 500,000 to 600,000 tonnes of wheat stocked at the farm level may have been damaged or lost to the floods.

The losses of wheat at the household level could also have a negative impact on planting of the 2010/11 *Rabi* season, which typically begins in October/November. The floods have resulted in the loss of seeds for planting; other agricultural inputs such as fertilizer; livestock; and savings. In addition, damage to irrigation facilities and wells could affect wheat production in the next *Rabi* season, as 90% of Pakistan's wheat crop is irrigated. The flooding will also negatively affect Pakistan's ability to export agricultural commodities. Requests from Punjabi government officials who wrote to the federal government asking for permission to export excess stockpiles of wheat, generated outrage from various aid agencies who claimed that the country's food situation remains "too precarious" to justify exporting crops. In September 2010, Pakistan's Economic Coordination Committee voted to ban all wheat exports in order to ensure that necessary food was not being sent outside of the country.[51]

FAO has established a response management team and is working in collaboration with the WFP and the International Fund for Agricultural

Development (IFAD) in flood-affected provinces as well as across Pakistan to provide technical assistance to meet three primary goals: meeting the 2010/11 *Rabi* wheat planting season, rehabilitating land and irrigation infrastructure, and saving livestock. FAO is providing critical agricultural inputs, such as seed, fertilizers, and farming tools to clear water-logged fields, to hundreds of thousands of farmers for the spring planting season. In October 2010, the FAO announced that it had reached over 532,000 families with agricultural inputs for the Rabi wheat season, 19,000 families with irrigation support, and over 163,000 families with animal feed and veterinary supplies.[52]

Global agricultural assistance efforts for Pakistan in response to the flooding have largely been underfunded.[53] On October 29, 2010, the U.S. Agency for International Development (USAID) committed $16 million to FAO to provide agricultural inputs to flood-affected families in Balochistan Province.[54] The funding complements similar programs in KPk and Punjab provinces and will provide wheat seed, vegetable seed, and fertilizer packages to about 680,000 family members—particularly women—and will also provide veterinary care and fodder to about 55,000 animals.[55]

Livestock

The floods have affected the most densely populated livestock areas in Pakistan. The national livestock population was estimated at 217 million animals in 2006, including cattle, buffalo, sheep, goats, donkeys, and poultry. Estimates show that over 1.2 million livestock and 6 million poultry have perished[56] In KPk alone, a rapid agricultural sector assessment by FAO found that over 67,000 head of small and large livestock have died, and almost 600,000 are suffering from malnourishment and sickness. In addition, poultry losses have been estimated in the millions, with the entire poultry stock wiped out in some areas. Many animals died because they had to be left behind when people were rescued by the Pakistani military and other rescue services.

FAO stated that "millions of surviving animals are now facing severe feed shortages, threatening generations of Pakistan's livestock,"[57] and that one of its primary priorities is maintaining and keeping healthy the surviving population of livestock. Getting feed and veterinary services to those in need continues to be a major challenge as supplies of animal feed such as straw and forage is in limited supply, and transportation of goods and services is severely limited due to considerable damage to critical infrastructure. FAO is working to reach tens of thousands of households with supplementary feed and veterinary support.

Economic Implications of the Floods

As the flood waters continue to recede, various entities are disclosing preliminary assessments of the economic impact of the disaster. An August estimate by economic advisors to Pakistan's Ministry of Finance reportedly set the flood losses at $12-15 billion.[58] The Asian Development Banks (ADB) and the World Bank (WB) released on October 14, 2010, a joint estimate of the flood damage at $9.7 billion.[59]

The broader short-term economic effects of the flooding are already being felt. The floods have sent food prices soaring, exacerbating fears of a new round of high inflation. In addition, greater government expenditures on the emergency relief effort are almost certain to widen the fiscal deficit. Reportedly, the United Nations has estimated that the floods will increase Pakistan's fiscal expenditures by $4 billion in fiscal year 2010/2011.[60] A preliminary assessment submitted to Pakistan's Minister of Finance has estimated that there may be zero real GDP growth and 25% inflation in the current fiscal year, compared to the IMF targets of 4.5% real GDP growth and 9.5% inflation.[61] Pakistan's consumer price index (CPI) increased by 15.7% in September 2010. Other estimates predict real GDP growth of about 2%-3% for the current fiscal year, compared to 4.1% in the previous fiscal year.[62]

Pakistan may also face another balance of payments crisis like it did in the autumn of 2008.[63] Agricultural production not only plays a critical role for Pakistan's domestic economy, it also is important for its exports. Pakistan's rice exports in 2009 were worth $1.8 billion, accounting for 10% of its exports. Cotton and cotton yarn exports in 2009 totaled $3.2 billion, or 18% of Pakistan's total exports. Cotton is also vital for Pakistan's other leading exports—clothing and other textile articles—which together were worth $5.8 billion, or 33% of total exports. A slow recovery in agricultural production will adversely effect its balance of trade and potentially contribute to a balance of payments crisis. Pakistan held $17.1 billion in foreign exchange reserves as of October 15, 2010.

The long-term economic effects are more difficult to assess. According to one source, repairing damaged infrastructure, including countless destroyed roads, bridges, and dams, could cost up to $15 billion.[64] Also, as discussed earlier, the World Bank and ADB have estimated $9.7 billion in damages in a preliminary report. Critical factors that will influence the speed of Pakistan's economic progress over the next few years are the pace of the recovery of the agricultural sector, the status of the millions of displaced people, and the level of international support for the relief and reconstruction assistance.

The ADB has offered a loan of $2 billion and the World Bank has offered a loan of $1 billion to be used primarily in the reconstruction of Pakistan's infrastructure damaged by the floods. The Friends of Democratic Pakistan (FODP) met in October 2010, in Brussels to discuss potential ways to bolster the Pakistani economy in the wake of flooding, among other things.[65] Some proposals included providing trade concessions on selected products, opening up selected EU markets, and encouraging Pakistan to increase its tax base.[66] The FODP also pressed the Pakistan government to continue to implement economic reforms, including broadening its tax base.

Pakistan's Finance Minister Abdul Hafeez Sheikh was in Washington the week of August 23-27, 2010, to meet with the IMF as part of the fifth review of the IMF loans. On November 20, 2008, the IMF's Executive Board approved a SDR 5.169 billion (at the time, the equivalent of $7.65 billion)[67] stand-by arrangement for Pakistan to avert a balance of payments crisis. The IMF subsequently approved an additional SDR 2.067 billion ($3.22 billion) in assistance on July 27, 2009. These loans were to be paid out in seven installments (called "tranches"), ending in November 2010, following reviews of Pakistan's economic progress. As part of the loan agreement, Pakistan promised to make a number of changes in fiscal and monetary policy (commonly known as "conditionalities"), including gradually lowering its fiscal deficit as a percentage of GDP. The agreement also stipulated that Pakistan would have to repay the loans starting in 2012.

Prior to the floods, Pakistan had failed to meet several of the IMF conditionalities, and the IMF was reportedly applying pressure on Pakistan to meet its commitments. Pakistan's fiscal deficit has increased to 6.2% of GDP, a level exceeding the IMF target of 5.1% of GDP. The cost of the emergency relief efforts are likely to make it impossible to meet the IMF target. Pakistan reportedly asked the IMF to relax this and other conditionalities during the August meetings in Washington. It also requested the release of the remaining balance of the IMF loans, worth about $2.6 billion. While the IMF decided to delay the release of the next tranche, on September 16, it did announce a $451 million emergency loan to help Pakistan deal with the aftermath of the flooding.

In early November, the IMF and Pakistan government again met to discuss the economic situation in Pakistan. While Pakistan reportedly pressed for the release of the remaining tranches, the IMF focused on the need for additional economic reforms. For the IMF, the key issues are: the implementation of reformed general sales tax (RGST), which Pakistan was supposed to have enacted in July 2010; eliminate the remaining subsidies for electricity; and reducing the fiscal deficit. Pakistan has until December 13, 2010, to make the necessary economic reforms or lose access to the remaining $1.7 billion of the stand-by arrangement.

The Pakistan government has made some efforts to respond to the economic impact of the floods and the demands of the IMF. Pakistan's National Oversight Disaster Management Council has developed a $7.7 billion reconstruction plan. However, according to the plan, the Pakistan government would only provide $3.8 billion of the funds and seek $3.4 billion in international assistance; the remaining $500 million would supposedly come from "other sources." Pakistan's fiscal deficit was also worsened by the 25% increase in Pakistan's defense budget announced on September 23, 2010.[68] The increased military expenditures were to be used for major military operations against the Taliban in the tribal areas bordering Afghanistan. Part of the increased military outlays are to be financed by cuts in other areas of the federal budget. In addition, the Pakistan government has reportedly promised the IMF to raise the electricity rates by 2.2% per month until July 2011—a total increase of 17.6%.[69] On November 15, 2010, the IMF denied that such a deal had been struck with the Pakistan government.[70]

Another economic problem looming on the horizon is the cost of servicing the IMF loans and other debt incurred since 2008. Pakistan's current outstanding debt is $55.5 billion, which is 56% of GDP. Under the current agreement, repayments on the IMF loan will cost the Pakistan government over $3 billion per year starting in fiscal year 2012/2013. There is concern that the cost of servicing its debts may undermine Pakistan's economic recovery.

Political-Security Implications of the Flood[71]

The Obama Administration's strategy toward Pakistan, and the broad goals of the Enhanced Partnership With Pakistan Act of 2009 (P.L. 111-73, also known as the Kerry-Lugar-Berman act), are likely to be significantly frustrated by the worst natural disaster in Pakistan's 63-year history. Central to the provisions of P.L. 111-73 was the authorization of $7.5 billion in nonmilitary U.S. assistance to Pakistan for the period FY2010-FY2014, a major increase in economic and development aid aimed at bolstering Pakistan's institutions and infrastructure, as well as demonstrating a long-term U.S. commitment to partnership with that country. Many observers now consider it likely that a portion of any funds appropriated in the future will be diverted to recovery efforts and could thus be unavailable for the types of capacity-building programs and projects originally envisaged.

In the nearer-term, U.S. interests in countering Islamist militancy in the region, and strengthening Pakistan's democratic institutions are under greater threat due to the chaos and destruction caused by widespread flooding there.[72] While ongoing U.S. and NATO military operations in neighboring Afghanistan are relatively unaffected, the circumstances inside Pakistan appear to be providing time and space for both Afghan and Pakistani Taliban militants, as well as numerous other assorted indigenous Islamist extremist groups, to regroup. Moreover, by undertaking disaster relief efforts in areas poorly served by the government, Islamist charities linked to such groups could boost recruitment and improve their standing with affected residents. The U.S. government is actively seeking to counter this possibility through its own ground response, with officials in Washington viewing the catastrophe as an opportunity to demonstrate America's commitment to and friendship with the Pakistan people. After surveying the affected areas in person, the Chairman of the Senate Foreign Relations Committee, Senator John Kerry, declared that helping Pakistan "is in all of our interests."[73]

On the political front, an already weak civilian government in Islamabad is criticized for what has been viewed as a slow and woefully insufficient response, further damaging its credibility.[74] In contrast, the Pakistani military, widely viewed as the country's most relatively capable institution, has benefited in the public eye following its energetic relief efforts in affected areas. For some observers, this dynamic could threaten recent progress toward democracy in a country that has seen four military coups in the past 42 years.

Pakistan-U.S. Relations, Militant Groups, and Public Sentiment

Potential to Improve U.S. Standing in Pakistan

Hopes abide that the rapid and major disaster relief assistance being provided by the United States may contribute to lessening widespread and long-held anti-American sentiment in Pakistan. In this respect, the tragedy could also be seen as an opportunity to narrow the gaping "trust gap" in U.S.-Pakistan relations. The U.S. Special Representative for Afghanistan and Pakistan, Ambassador Richard Holbrooke, has been explicit in seeking this goal.[75] By some accounts, this is taking place, although the evidence remains anecdotal.[76] Yet there is also evidence that Pakistanis are largely unaware of U.S. largesse in the recent past and do not report seeing tangible benefits from same. According to one recent opinion survey, about half of Pakistani respondents believe the United States gives little or no assistance to their country, suggesting that public relations gains from aid may be fleeting, at best.[77] Moreover, while the U.S. government has committed $200 million to date, private contributions have been quite small relative to those given for other recent natural disasters, potentially fueling Pakistani resentment and distrust.[78]

Islamist Extremists Groups and Disaster Relief

Some extremist-run charities have undertaken relief efforts in areas where government aid has been lacking. By providing food, shelter, and other benefits to desperate victims, such organizations may win sympathy and even (additional) future support from affected residents. Of potential concern, especially from a U.S. perspective, are the activities of the Falah-i-Insaniat and Jamaat-ud-Dawa. (JuD). These are the charity and political wings, respectively, of Lashkar-eTaiba (LeT), a U.S.-designated Foreign Terrorist Organization (FTO) held responsible for the Mumbai attack of November 2008, as well as many other terrorist activities.[79] When asked about the likelihood of extremist groups gaining strength and influence through their own relief activities, Secretary of State Clinton said she did not think it was great and suggested that such reports were being overstated.[80]

Another U.S.-designated FTO, the Tehrik-i-Taliban Pakistan (TTP, also known as the Pakistani Taliban) denounced U.S. assistance and called for a boycott of all Western aid.[81] Although the TTP has threatened to attack Western aid workers in the country since the floods began, no such attacks are known to have taken place, and the U.S. military is taking significant precautions to ensure the safety of own its forces, as well as for other U.S. aid workers in-country. In late August, a senior U.S. military officer in Pakistan stated that he had "seen no security threat whatsoever" to U.S. personnel in Pakistan in the preceding three weeks, and he lauded the Pakistani military for its provision of force protection.[82]

Domestic Political Effects

Pakistan's particular circumstances—poor government capacity, struggling economy, sagging infrastructure, multiple armed insurgencies, and all manner of regional, sectarian, and class divisions—leave it ripe for potentially major upheaval as a result of the floods. In the view of some analysts, the aggregated pressures are so great that the current national government is seen as unlikely to survive and complete its five-year term set to end in the spring of 2013.[83] Some more alarmist observers have even suggested that the Pakistani state itself could collapse, although most believe this is a remote possibility.[84]

Status of the Civilian Government

The current national coalition government was seated in the spring of 2008 after relatively free and fair elections. It is led by the Pakistan People's Party (PPP) of President Asif Zardari and Prime Minister Yousaf Raza Gilani. After flooding began in late July, President Zardari came under intense criticism for going ahead with a planned state visit to Europe even as the scope of the devastation was coming clear. His early August travel, which included costly stays at luxury hotels, hugely exacerbated a pre-existing perception of the President as detached from the human suffering taking place in his own country.[85] Zardari contended that he had used the trip to "mobilize foreign assistance" and "had to choose substance over symbolism."[86] Nevertheless, Pakistanis appear to feel significant anger toward politicians at all levels of government, and this sense of abandonment in time of need could significantly weaken the standing of democratic forces in the country.[87] One senior Pakistani analyst saw both the federal and provincial government responses as representative of the broader "abysmal failure" of the country's political class. He predicts "catastrophic consequences" for the PPP-led national government if relief work is not undertaken efficiently and with transparency.[88] In a country where corruption and political patronage are rife, angry accusations have arisen that officials are manipulating the distribution of relief resources to their political allies.[89]

Internal Administrative Issues

The internal political effects of the floods remain unclear to date, but are likely to be quite serious. There have been signs that Pakistan's already fraught center-province and interprovincial tensions are being exacerbated by economic hardship and competition for resources. For example, in the densely populated Punjab province, home to more than half of Pakistan's total population, the provincial government of former Prime Minister Nawaz Sharif's Pakistan Muslim LeagueNawaz (PNL-N, which represents the main opposition at the national level) has tussled with the federal government over aid disbursement issues. The PML-N may seek to take advantage of anti-government sentiments to make political inroads in the Sindh province, home to the PPP's core votebank.[90] Meanwhile, there have been credible allegations that rich landowners, mostly in Sindh, allowed levees to burst so that floodwaters would be diverted to unprotected areas, sparing their crops. These actions may have had a negative impact on the neighboring Baluchistan province.[91]

Role and Standing of the Pakistani Military

Pakistan's military has always played a key role in the country's governance; the country has been ruled directly by the military for more than half of its existence as an independent state. The ultimately unpopular rule of President-General Pervez Musharraf, who seized power though unconstitutional means in 1999 and left office in 2008, was harmful to the military's public image, which typically rates as the most positive among Pakistani institutions. Since his 2007 ascension to the pinnacle of Pakistani military rank, current Chief of Army Staff General Ashfaq Pervez Kayani has worked with considerable success to restore the army's prestige while at the same time withdrawing it from any overt role in governance. With his organization taking the lead in providing flood relief, Kayani's influence has only grown, and the military's Rawalpindi headquarters is again seen by many as the sole effective locus of state power in Pakistan.[92] The implications are worrisome for proponents of representative government who wish to see Pakistan's democratic institutions strengthened further. The U.S. government explicitly supports Pakistani democracy. When asked about recent controversial statements, a U.S. State Department spokesman said, "Pakistan has a civilian government and we think it is the best form of government to take."[93] More recently, Prime Minister Gilani assured the nation that a military takeover was out of the question.[94]

Counterinsurgency Operations

Almost immediately upon incoming reports of this latest natural disaster in Pakistan, fears arose in Washington, Islamabad, and other world capitals that resulting chaos and destruction would open up space for Pakistan's numerous Islamist militant groups to regroup and perhaps even recover from losses suffered from recent, damaging battles with the Pakistani military. There have already been some reports of militants exploiting the situation with attacks on government security targets, and devastating suicide bomb attacks against minority Shia demonstrate that the threat posed by terrorist groups has not subsided as a result of the flooding. Since early August, some 60,000 Pakistani army troops were diverted to disaster relief efforts. In mid-month, a spokesman for the Pakistani military asserted that their relief activities would "have no impact on our fight against militants."[95]

President Zardari has warned that "negative forces" could see a boost in recruitment as a result of the catastrophe. In a similar vein, Foreign Minister Qureshi expressed worries that, in the absence of sufficient government intervention, food rioting could lead to further violence "being exploited by people who are known." Provincial officials have issued similar warnings.[96] Some observers believe such warnings are overblown and may reflect the desire of Pakistani officials to attract greater foreign assistance.

In recognition of the overwhelming nature of the flooding, U.S. officials apparently have refrained from pressing the Pakistani government and army to go ahead with anticipated military operations in the FATA—especially against the Afghan Taliban-aligned Haqqani network in North Waziristan—as they had been throughout 2010.[97] In August, the deputy commander of the Office of the U.S. Defense Representative in Pakistan expressed confidence that the Pakistanis will "continue to wage a dedicated, committed struggle against violent extremism." More recently, the commander himself said the numbers of Pakistani troops dedicated to counterinsurgency operations was unchanged and the focus of the their leadership had "not wavered." It is notable that the militants also have suffered setbacks: in the words of the senior-most U.S. military officer, "I think [the flooding] delays everybody's timetable," including "the enemy's timetable."[98]

Despite these official U.S. and Pakistani expressions of confidence and determination, most independent observers contend that counterinsurgency efforts in the country now face major and obvious setbacks, not only due to the need to divert resources to relief efforts, but also because so much infrastructure has been damaged or destroyed.[99] Military gains made in the past 18 months are thus in jeopardy of being lost as a result of the flooding. In the Swat Valley of the KPk province, which the Pakistani military had largely cleared of anti-government militants in 2009, electricity has been widely unavailable and most key roads, bridges, and other infrastructure are unusable.[100] Likewise, although TTP bases in South Waziristan recently were cleared through major military offensives, efforts to resettle hundreds of thousands of citizens displaced by the fighting there have been placed on hold.[101] Beyond these Pashtun-majority northwestern regions of the country, the flooding may also contribute to increased conflict and instability in the Pakistani heartland, where the LeT and other Sunni militant groups are based. Moreover, by creating a flow of hundreds of thousands of internally displaced persons into the southern megacity of Karachi, the floods have placed even more pressures where interethnic tensions were already high.[102]

Some press reports have indicated that the main supply lines for U.S. and NATO forces in Afghanistan, which run from the Karachi port north to through Quetta or Islamabad, were significantly disrupted by the flooding. The inundation has forced trucks to take longer routes than usual, delaying schedules, but a spokesman for the U.S. military's Transportation Command said the slower delivery was not meaningfully affecting operations in Afghanistan. While the immediate affects on US/NATO operations are thus minor, Pentagon officials are concerned that secondary and tertiary effects impacting regional stability could be seen in the medium and longer-term.[103] The Afghan military has provided some resources to Pakistani flood relief, potentially improving Pakistan-Afghan relations.

Context of Pakistan's Flood Risk

Pakistan has experienced severe and dangerous flooding in the recent past.[104] One source indicates that serious flooding has affected what is now Pakistan at least 12 times since 1928.[105] This is the fifth flood in Pakistan affecting nearly 5 million or more people in less than 40 years.[106]

Extreme precipitation is the immediate cause of flooding, and is part of weather extremes spawned during the dominant climate condition for Pakistan—the Asian monsoon system.[107] Rivers transporting floodwater from the north can pose flood risks to areas downstream. While extreme precipitation is the flood hazard, development patterns and infrastructure choices affect the number of people and structures and the infrastructure at risk. The relative frequency of significant flooding in Pakistan raises questions about if and how efforts to rebuild can improve the nation's resiliency to future extreme weather events and how international experience and aid can inform resilient rebuilding.

The current flooding in Pakistan is likely to prove the worst flooding disaster in the country's history, as far as the number of citizens affected, and the cumulative damage to the region's economy. It is not yet clear to what extent the flooding itself, in terms of the flow, inundation duration, and land area affected exceeds previous flooding episodes, in part because of the paucity of instrumental data and lack of detailed analyses of recent past and paleo-flooding events in the region. Forthcoming studies are likely to shed light on whether the current flood fits within the historical range of extreme flooding events, or truly represents an outlier compared to the 1,000 year record of high rainfall events captured in tree-rings.[108]

Climate Change and Flooding

A related issue is whether climate change during the last half-century contributed to the extent and severity of the current flooding. The 2007 Intergovernmental Panel on Climate Change (IPCC) report stated that scientists project more frequent and more intense weather events due to global warming.[109] However, the report and other scientists contend that a longer time range is needed to attribute single weather events, such as the 2010 floods in Pakistan, to global warming. The IPCC report also stated that an increase in precipitation is projected in the Asian monsoon associated with man-made global warming. The report points out, however, that the role of aerosols in general, and carbon aerosols in particular, complicates the nature of monsoon precipitation, especially Asian monsoon.[110] Detecting a global-warming induced fingerprint in the current flooding may prove elusive until more detailed and focused scientific studies of the nature of the current flooding are completed. Even if global warming were identified as a contributing factor to the severity of the current flooding, it may not be possible to equate the global warming "signal" to an increase in deaths, injuries, and economic losses because changes in population demographics, infrastructure, and other factors also contribute to vulnerability to weather extremes. Disentangling climate change effects from demographic factors contributing to economic losses has been notoriously difficult for natural disasters such as floods and tropical cyclones in the United States.[111] Therefore, although the monsoon's floodwaters appear broadly consistent with predicted climate change impacts for the region, it is currently not possible to attribute direct causality.

OBSERVATIONS

The flooding in Pakistan is expected to have significant short-term and potential long-term effects on the country. These effects will be exacerbated by existing problems in Pakistan that range from poverty and depleted infrastructure, among other things. The short-term effects of flooding are steadily being realized as displaced people start returning to their homes in northern areas where floodwaters have receded, and government officials and aid workers begin reporting issues from all affected areas. The full extent of the long-term effects of the flooding are likely to present daunting challenges to the country. The long-term effects are likely to manifest themselves in two ways that have significance to the United States and Congress. One aspect is the humanitarian toll that is likely to emerge from displaced people, disease, food security, and an economic decline. Another aspect is the strategic concerns that could result from a weakened government, and a dissatisfied and disenfranchised population. The strategic implications involve U.S. interests in countering Islamic militancy in the region, strengthening democracy in Pakistan, and fighting the war in Afghanistan.

End Notes

[1] United Nations Office for the Coordination of Humanitarian Affairs, *Pakistan—Monsoon Floods*, Situation Report #23, September 9, 2010. Hereafter referred to as *OCHA Situation Sept. 9 Report.*

[2] European Union, *Factsheets on Pakistan*, European Union, Memo, Brussels, October 1, 2010.

[3] The United Nations Office for the Coordination of Humanitarian Affairs, Pakistan Floods Affect Millions from North to South, August 9, 2010.

[4] "In Pakistan, Taliban Hint at Attacks on Relief Workers," *The New York Times,* August 26, 2010.

[5] For additional information on broader dynamics related to security and the environment in Pakistan see CRS Report R41358, *Security and the Environment in Pakistan.*

[6] Pew Research Center, *Pakistan: Growing Concerns About Extremism, Continuing Discontent with U.S.* Public Opinion Survey, August 13, 2009. See at http://pewresearch.org/pubs/1312/pakistani-public-opinion.

[7] For example, bills introduced in the 111th Congress would provide duty free treatment for certain goods, particularly textiles and some clothing, from designated reconstruction opportunity zones in Pakistan and Afghanistan. Textiles represent roughly half of Pakistan's exports.

[8] Office of the Coordination of Humanitarian Affairs, *Pakistan Humanitarian Bulletin*, United Nations, Issue 5/4, November 2010.

[9] United Nations, *Pakistan: Floods Relief and Early Recovery Response Plan*, United Nations, November 2010, http://pakresponse.info/LinkClick.aspx?fileticket=47teGm9PeB8%3d&tabid=93&mid=676.

[10] International Crisis Group, *Pakistan: The Worsening IDP Crisis*, International Crisis Group, Asia Briefing no. 11, Brussels, Belgium, September 16, 2010.

[11] Talk given by General Jehangir Karamat, *Responding to the Historic Floods in Pakistan: Political and Security Considerations*, Brookings Institution, Washington, DC, September 1, 2010.

[12] USAID Fact Sheet #6 Fiscal Year 2011, November 12, 2010.

[13] Ibid.

[14] Xinhua General News Service, September 9, 2010.

[15] Funding levels are from Factsheets on Pakistan http://www.reliefweb.int/rw/RWFiles2010.nsf/FilesByRWDocUnidFilename/MUMA-89Z5BL-full_report.pdf/$File/full_report.pdf.

[16] Embassy of Japan in Pakistan, Press Release No: JPNEMPAK 10-038, September 8, 2010, http://www.pk.embjapan.go.jp/InfoJapan/Press%202010/September/JPNEMPAK%20010-38html.

[17] The Canadian International Development Agency, News Release, *Government of Canada Supports Agricultural Recovery in Pakistan,* October 15, 2010.

[18] United Nations, *Pakistan: Floods Relief and Early Recovery Response Plan*, United Nations, November 2010, http://pakresponse.info/LinkClick.aspx?fileticket=47teGm9PeB8%3d&tabid=93&mid=676.

[19] As of November 2010, approximately 39% of the revised appeal was covered.

[20] WHO, Flooding and Communicable Diseases Fact Sheet: Risk assessment and Preventive Measures, at http://www.who.int/hac/techguidance/ems/flood_cds/en/.

[21] Pakistan Health Cluster, "Floods in Pakistan, Bulletin No 19," September 28, 2010, at http://www.whopak.org/idps/ documents/bulletins/Health%20Cluster%20Bulletin%20No%2019-Final.pdf, "Floods in Pakistan, Bulletin No 12," August 16, 2010, at http://www.who.int/hac/crises/pak/sitreps/16august2010/en/index.html.

[22] Meriwether Beatty, "Under Water: Women's Access to Reproductive Health Care in Pakistan," Global health Magazine, August 25, 2010, at http://www.globalhealthmagainze.com/gust_blog_under_water/.

[23] Kaiser Family Foundation, "At Least 3.5M Pakistanis Have No Access To Clean Water, Raising Risk For Waterborne Diseases," August 26, 2010, at http://globalhealth. kff.org/Daily-Reports/2010/August/26/GH-082610- Pakistan.aspx.

[24] World Health Orgnaization, "Cholera in Pakistan," October 26, 2010, at http://www.who.int/csr/don/2010_10_25/ en/.

[25] Pakistan Health Cluster, "Weekly Epidemiological Bulletin: Flood Response in Pakistan," Volume 1, Issue 7, October 4, 2010, at http://www.whopak.org/idps/documents/wmmrs/wmmr_flood_w39_2010.pdf.

[26] United Nations, Office for the Coordination of Humanitarian Affairs, "Pakistan: 3.5 Million Children at Risk of Deadly Disease," press release, August 16, 2010, at http://www.reliefweb.int/rw/rwb.nsf/db900sid/VVOS-88DPRA? OpenDocument&query=children&emid=FL-2010-000141-PAK.

[27] WHO, Communicable Disease Risk Among Populations Affected by Flooding in Pakistan, August 2010, at http://www.who.int/diseasecontrol_emergencies/publications/Pakistan_floods2010_08_04.p df.

[28] Pakistan Health Cluster, "Floods in Pakistan, Bulletin No 19," September 28, 2010, at http://www.whopak.org/idps/ documents/bulletins/Health%20Cluster%20Bulletin% 20No% 2019-Final.pdf; World Health Organization, "Crimean-Congo haemorrhagic fever (CCHF) and Dengue in Pakistan," October 25, 2010, at http://www.who.int/csr/don/ 2010_10_25a/en/index.html.

[29] International Crisis Group, *Pakistan: The Worsening IDP Crisis*, International Crisis Group, Asia Briefing no. 11, Brussels, Belgium, September 16, 2010.

[30] Carlotta Gall, "After Flood, Pakistani Officials Try to Quell Anger," *International Herald Tribune*, September 7, 2010, p. 3.

[31] Elizabeth Ferris, "Natural Disasters and Human Rights: Comparing Responses to Haiti and Pakistan," Presentation at Center for Human Rights and International Justice, Boston College, November 3, 2010.

[32] Ibid.

[33] Humanitarian relief sectors or clusters are typically established by the United Nations during humanitarian crises to help coordinate partners, prioritize resources, and facilitate planning.

[34] *30 Earthquakes, 28 Lessons Learned,* OneWorld.net, January 21, 2010, http://us.oneworld.net/article/368623-30-earthquakes-28-lessons-learned and *The Haiti castrophe: lessons learned from previous operations*, by Ben Ramalingam, Overseas Development Institute (ODI) Blog January 19, 2010, http://blogs.odi.org.uk/blogs/main/archive/2010/01/19/lessons_for_haiti_aid_response.aspx.

[35] United National Institute for Training and Research, The Pakistan Floods of 2010: Flood Analysis in Pakistan Based on Time Series of Satellite Data Recorded from 28 July to 16 September 2010, September 20, 2010, http://unosatmaps.web.cern.ch/unosat-maps/PK/FL20100802PAK/UNOSAT_PAK_FL2010_EarlyRecoveryOverview_v2_HR.pdf; Carlotta Gall, "Pakistan flood sets back years of gains on infrastructure," New York Times, August 26, 2010; OCHA Situation Aug 27 Report. Also, more than 6,000 schools are being used as emergency shelters.

[36] Asian Development Bank, "ADB-World Bank Assess Pakistan Flood Damage at $9.7," press release, October 14, 2010, http://www.adb.org/Media/Articles/2010/13363-pakistan-flooding-assesments/ADB-WB-pakistanassessment.pdf.

[37] Ibid.

[38] Michael J. Hicks and Mark L. Burton, *Preliminary Damage Estimates for Pakistani Flood Events, 2010*, Center for Business and Economic Research, Ball State University, August 2010, http://cber.iweb.bsu.edu/research/ PakistanFlood.pdf. The analysis included primarily commercial and residential building losses and rail and highway impacts.

[39] Nathanial Gronewold, "After the Pakistan Deluge, Blackouts Spread," New York Times. October 12, 2010. http://www.nytimes.com/cwire/2010/10/13/13climatewire-after-the-deluge-blackouts-spread-15812.html.

[40] Energy Information Administration, "International Energy Statistics," U.S. Department of Energy. http://tonto.eia.doe.gov/cfapps/ipdbproject/IEDIndex3.cfm?tid=2&pid=2&aid=7. Latest figures are for 2007.

[41] Khalid Mustafa, "Deluge impact: power system still deprived of 2,796 MW," The News (International), Aug 27, 2010, http://www.thenews.com.pk/27-08-2010/National/1518.htm.

[42] *Natalie Obiko Pearson,* "Pakistan Flood Shuts Biggest Refinery, Cuts Natural Gas Output," Bloomberg Businessweek, August 11, 2010. http://www.businessweek.com/news/2010-08-11/pakistan-flood-shuts-biggestrefinery-cuts-natural-gas-output.html.

[43] "Flood victims exempted from electricity bills," Daily Times (Pakistan), Setpember 23, 2010. http://www.dailytimes.com.pk/default.asp?page=2010%5C09%5C23%5Cstory_23-9-2010_pg7_13

[44] Khurrum Anis, "Pakistan's Biggest Refinery Resumes Production After Floods," Bloomberg, September 20, 2010. http://www.businessweek.com/news/2010-09-20/pakistan-s-biggest-refinery-resumes-production-after-floods.html

[45] The Pakistani government has been increasing electricity tariff rates, a politically unpopular measure, in order to reduce subsidies and spur energy investment. These adjustments are part of Pakistan's program with International Monetary Fund and has the support of other multilateral institutions and the U.S. Department of State. Note that flood affected households have been exempted from electricity bills according to the Minster for Water and Power, Raja Pervaiz Ashraf (see Daily Times, 2010 above).

[46] Following the U.S. Congress's passage of the Enhanced Partnership with Pakistan Act of 2009 (P.L. 111-73), the U.S. Department of State launched the Pakistan Signature Energy Program. See U.S. Department of State, "The United States Announces Phase II of the Signature Energy Program for Pakistan," press release, July 19, 2010, http://www.state.gov/r/pa/prs/ps/2010/07/144820.htm, and "The United States Announces

the Pakistan Signature Energy Program," press release, October 28, 2009, http://www.state.gov/r/pa/prs/ps/2009/oct/130995.htm.

[47] International Fund for Agricultural Development (IFAD), *Islamic Republic of Pakistan: Country strategic opportunities programme*, Rome, March 25, 2009, http://www.ifad.org/gbdocs/eb/96/e/EB-2009-96-R-9.pdf.

[48] Rabi crops, such as wheat, barley, maize, peas, oats, and mustard, are typically planted in winter from October to December and harvested in spring from April to June; Kharif crops, such as millet, rice, corn, pulses, cotton, and soybean, are typically planted with the onset of the monsoon in different parts of the country, often between July and August, and then are harvested in the fall from September to October. In between the *Rabi* and the *Kharif* seasons there can be a short season during the summer months known as the *Zaid* season.

[49] FAO, *Pakistan Flooding: Executive Brief,* October 5, 2010, http://www.fao.org/fileadmin/templates/tc/tce/pdf/Pakistan_Floods_Executive_Brief_05_October_2010.pdf.

[50] FAOSTAT estimates Pakistan's total arable land to be about 23 million hectares.

[51] See http://www.criticalthreats.org/pakistan-security-brief/pakistan-security-brief-%E2%80%93-october-27-2010

[52] FAO, op. cit.

[53] As of September 2010, FAO had received about USD$46 million in donor contributions, out of an estimated total need of USD$107 million for agricultural assistance. For more information about FAO's component of the overall Revised Pakistan Floods Emergency Response Plan, see http://www.fao.org/fileadmin/templates/tc/tce/pdf/ PAKISTAN_REVISED_FLOODS_EMERGENCY_RESPONSE_PLAN_.pdf.

[54] See USAID press release: http://www.usaid.gov/pk/newsroom/news/disaster/101029.html

[55] See USAID, http://www.usaid.gov/pk/downloads/da/10.29.10-USAID-DCHAPakistan Floods FactSheet4-FY2011.pdf

[56] FAO, op. cit.

[57] See http://www.fao.org/news/story/en/item/44803/icode/.

[58] "Pakistan: Havoc Strains External Balances, cuts GDP," *Khaleej Times*, August 30, 2010.

[59] Asian Development Bank , "ABD-World Bank Assess Pakistan Flood Damage at $9.7 Billion," press release, October 14, 2010.

[60] Farhan Bokhari, "Pakistan Seeks $10bn Loan Restructure," *Financial Times*, August 19, 2010.

[61] "We're Looking at 0% GDP Growth and 25% Inflation," *The International News*, August 23, 2010. Pakistan's fiscal year runs from July 1 to June 30.

[62] Economist Intelligence Unit, "Pakistan Country Report," *Economist*, September 2010.

[63] For details of Pakistan's 2008 balance of payments crisis, see CRS Report RS22983, *Pakistan's Capital Crisis: Implications for U.S. Policy*, by Michael F. Martin and K. Alan Kronstadt.

[64] "Debts Pushing Pakistan to the Brink of Ruin," *The Independent*, August 22, 2010.

[65] On September 26, 2008, an ad-hoc group of interested nations and organizations met on the sidelines of a United Nations General Assembly session to discuss Pakistan balance of payments crisis. Initially called the "Friends of Pakistan," the group held its first ministerial meeting on April 17, 2009, in Tokyo. The group subsequently renamed itself the Friends of Democratic Pakistan (FODP).

[66] Friends of Democratic Pakistan, "Joint Communiqué ," press release, October 15, 2010.

[67] The SDR, or "special drawing right" is the official unit of account for the IMF. Its value is determined by a basket of 16 major currencies, including the U.S. dollar. As such, the exchange rate between the SDR and the U.S. dollar varies over time. Unless otherwise

noted, the U.S. dollar equivalents cited in the memo are valued using the exchange rate at the time of the event.

[68] "Pak Hikes Defence Outlay by Rs 110 Bn," *The Economic Times*, September 23, 2010.

[69] Sajid Chaudhry, "Power Tariff to Go Up by 17.6 Percent," *Daily Times*, November 2, 2010.

[70] Irshad Ansari, "We Didn't Ask Pakistan to Hike Power Tariff," *The Express Tribune*, November 15, 2010.

[71] See also CRS Report R41307, *Pakistan: Key Current Issues and Developments*, by K. Alan Kronstadt.

[72] "U.S. Strategy in Pakistan Upended by Floods," *New York Times*, August 19, 2010. Even before the flooding began, Pakistan was suffering from multiple armed insurgencies, major terrorist attacks on cities, sectarian violence, a teetering economy, lack of state services, and a largely impotent civilian administration (see, for example, Ahmed Rashid, "The Anarchic Republic of Pakistan," *National Interest*, August 24, 2010).

[73] John Kerry, "A Call for Help in Flood-Ravaged Pakistan" (op-ed), *Boston Globe*, August 29, 2010. See also David Ignatius, "The Case for Aiding Flooded Pakistan" (op-ed), *Washington Post*, August 29, 2010.

[74] Numerous anecdotal reports from the ground indicate that, even as of early September, there had been no notable government response across wide swaths of affected areas (CRS interviews with returning relief workers, September 15, 2010).

[75] "U.S. Offers Aid to Rescue Pakistanis and Reclaim Image," *New York Times*, August 15, 2010. See also Shuja Nawaz, "A Chance to Build Trust in Pakistan" (op-ed), *Washington Post*, August 12, 2010. There is some empirical evidence that such a dynamic occurred in the recent past: A Pew poll taken shortly before Pakistan's catastrophic October 2005 earthquake found only 23% of Pakistanis expressing a favorable view of the United States, the lowest percentage for any country surveyed. That percentage doubled to 46% in an ACNielsen poll taken after large-scale U.S. disaster relief efforts in earthquake-affected areas, with the great majority of Pakistanis indicating that their perceptions had been positively influenced by witnessing such efforts.

[76] "US Aid Winning Friends in Flood-Ravaged Pakistan," Associated Press, August 10, 2010; "Opinions of U.S. Rise Along With Airlifts," *Washington Post*, August 12, 2010; "In Pakistan, U.S. Hopes Aid Builds Goodwill," *USA Today*, September 2, 2010. A recent study on the effects of foreign aid following Pakistan's October 2005 earthquake found that many "hearts and minds" were, in fact, won by the infusion ("Study: Aid After 2005 Quake Won Trust in Pakistan," *Washington Post*, September 7, 2010).

[77] "U.S. Aid Buys Little Goodwill," *Washington Post*, August 24, 2010.

[78] "In Pakistan, A Flood of Aid and Mistrust," *Washington Post*, September 6, 2010; "Pakistan Floods: Could 'Donor Fatigue' Harm US-Pakistan Relations?," *Christian Science Monitor*, September 8, 2010.

[79] The JuD is reported to have distributed thousands of packages containing food, soap, and other personal effects, along with 3,000 rupees (about $35) in cash ("Islamist Charity Aims to Be Pakistani's Salvation," *Reuters*, September 3, 2010).

[80] See the State Department's August 19, 2010, transcript at http://www.state.gov/secretary/rm/2010/08/146100.htm.

[81] "Pakistani Taliban Urge Rejection of Foreign Flood Aid," *Agence France Presse*, August 11, 2010. Even a senior leader of a mainstream Islamist party, the Jamaat-e-Islami, labeled aid from the United States (and India) as "poison" for the Pakistani nation and people ("Pakistan Islamist Party Says US, India Funds 'Poison' for Flood Victims," BBC Monitoring South Asia, August 25, 2010).

[82] "Mullen Shares Concern for Aid Workers," Armed Forces Press Service, August 27, 2010; Brig. Gen. Michael Nagata quoted in the August 25, 2010, Pentagon transcript at http://www.defense.gov/transcripts/transcript.aspx? transcriptid=4677.

[83] "Floods in Pakistan Carry the Seeds of Upheaval," *New York Times*, September 6, 2010.

[84] "Pakistan Flood Crisis Raises Fears of Country's Collapse," McClatchy News, August 13, 2010.

[85] In a comparison to President G.W. Bush's response to hurricane devastation on the U.S. Gulf Coast in 2005, one critic called the floods "Zardari's Katrina" (Fatima Bhutto, "Zardari's Katrina," *Foreign Policy* (online), August 4, 2010).

[86] Asif Ali Zardari, "Pakistan's Project of Renewal" (op-ed), *Wall Street Journal*, August 11, 2010.

[87] "Will Slow Response to Pakistan Floods Threaten Democracy?," *Christian Science Monitor*, August 12, 2010.

[88] Hasan-Askari Rizvi, "Is There a Way Out?" (op-ed), *Daily Times* (Lahore), August 15, 2010. See also Mohsin Khan and Shuja Nawaz, "Pakistan: Assessing the Tragedy," *New Atlanticist* (online), August 26, 2010.

[89] "Floods in Pakistan Carry the Seeds of Upheaval," *New York Times*, September 6, 2010.

[90] "National Flood Relief Strategy" (editorial), *Daily Times* (Lahore), September 6, 2010; "Pakistan After the Floods: A Bird's Eye View," IHS Global Insight, September 9, 2010.

[91] "Pakistan Landlords 'Diverted Flood Water,'" BBC News, September 2, 2010. A former Pakistani Prime Minister from Baluchistan accused members of the government of purposefully redirecting floodwaters in his home province, saying the actions reflected Islamabad's long-standing disregard for Baloch well-being and would prove costly to it in terms of public sentiment ("Floods in Pakistan Carry the Seeds of Upheaval," *New York Times*, September 6, 2010).

[92] "Flood Gives Renewed Clout to Powerful Pakistan Army," *Reuters*, September 6, 2010. The Army reports that it is providing more than half of all relief supplies, about three times the amount being provided by the rest of the government ("Pakistan Army Seeking Flood Relief Without Dependency on Aid, Mullen Told," *Bloomberg News*, September 2, 2010).

[93] "Altaf Asks 'Patriotic Generals' to Act," *News* (Karachi), August 23, 2010; Imtiaz Alam, "Altaf Hussain's Martial Law Balloon" (op-ed), *Daily Times* (Lahore), August 29, 2010; State Department's August 26, 2010, press briefing at http://www.state.gov/r/pa/prs/dpb/2010/08/146322.htm.

[94] "Army Won't Take Over: PM," *Daily Times* (Lahore), September 10, 2010.

[95] "Pakistan Floods Could Give Taliban Time to Regroup," *Associated Press*, August 11, 2010; spokesman quoted in "Pakistan Says Won't Divert Forces From Militant Fight," *Reuters*, August 13, 2010. Recent air strikes on Pakistani Taliban targets in the Tirah Valley of the Khyber Pakhtunkhwa province suggest that a level of pressure on insurgents is continuing ("Pakistan Raids Kill 45 Militants, Family Members," *Reuters*, September 1, 2010).

[96] "Pakistan Leaders Says Militants Could Exploit Flood," *Associated Press*, August 19, 2010; Qureshi quoted in "Pakistan's Epic Flood" (editorial), *Wall Street Journal*, August 19, 2010; "Pakistan Says Militants Exploiting Flood Crisis," *Los Angeles Times*, August 20, 2010.

[97] "Pakistan Fight Stalls for U.S.," *Wall Street Journal*, August 13, 2010.

[98] Brig. Gen. Michael Nagata quoted in the August 25, 2010, Pentagon transcript at http://www.defense.gov/transcripts/ transcript.aspx?transcriptid=4677; Vice Adm. Michael LeFever quoted in the September 9, 2010, U.S. Embassy transcript at http://islamabad. usembassy.gov/pr-10090903.html; Joint Chiefs Chairman Adm. Mike Mullen quoted in

"Pakistan Army Seeking Flood Relief Without Dependency on Aid, Mullen Told," Bloomberg News , September 2, 2010.

[99] See, for example, "Torrent of Challenges for U.S. in Pakistan" (interview), Council on Foreign Relations, August 20, 2010. Pakistan's ambassador to the United States reportedly conceded that the military's resources are stretched thin by the disaster, meaning its counter-insurgency operations are being hampered ("Flooding Hurts Pakistan Terror Fight," Associated Press, September 8, 2010).

[100] "Flood Brings Chaos Back to Pakistan's Swat Valley," *New York Times*, August 19, 2010.

[101] "For Pakistanis, The Worst May Be Still to Come," *Washington Post*, August 18, 2010.

[102] Ethnic Pashtuns displaced by conflict in the FATA are now joined in large numbers by ethnic Sindhis from rural areas, posing a potential threat the established order of Karachi's dominant Urdu-speaking Mohajirs. Rioting and targeted killings are ongoing in Karachi; about 1,000 people have been killed in ethnic strife in Karachi in 2010 to date ("Crush of Refugees Inflames Karachi," *Wall Street Journal*, August 26, 2010).

[103] "Pakistan Flooding Disrupts Afghan War Supplies," *New York Times*, August 24, 2010; CRS interview with Defense Department official, August 24, 2010.

[104] The topography, proximity to oceans, and other factors all create significant variability in precipitation across the Pakistani territory during the wet monsoon season. Parts of Pakistan, such as the desert region of the south and southeast, experience little precipitation relative to the eastern and central plains of the country, which receive the full effect of the monsoon rains.

[105] "Unnatural Disasters," *Financial Times*, August 4, 2010. Statistics for the flood illustrate that this is Pakistan's largest disaster in terms of the number of people affected in the last 60 years: http://www.reliefweb.int/rw/ fullmaps_sa.nsf/luFullMap/8A7B7152D23697D0C 125777B00411D87/$File/FL-2010-000141- PAK_0809_graph.pdf?OpenElement.

[106] Ibid.

[107] Defined by the National Weather Service as "a thermally driven wind arising from differential heating between a land mass and the adjacent ocean that reverses its direction seasonally." See http://www.nws.noaa.gov/glossary/.

[108] New scientific research is emerging that provides detailed information on the location, duration, and severity of drought and rainfall events in the region affected by the Asian monsoon, which includes India and Pakistan. In a recent study, tree ring records were examined to reconstruct dry and wet periods over the past 1,000 years for regions of monsoon Asia. Edward R. Cook et al., "Asian Monsoon Failure and Megadrought During the Last Millennium," *Science*, vol. 328 (April 23, 2010), pp. 486-489. Although the study focused on periods of drought (triggered when the monsoon "failed"), it noted that "Monsoon failures, megadroughts, and extreme flooding events have repeatedly affected the agrarian peoples of Asia over the past millennium."

[109] IPCC, "Contribution of Working Group I to the Fourth Assessment Report of the Intergovernmental Panel on Climate Change," Chapter 10, Global Climate Predictions, 2007.

[110] Ibid, Executive Summary.

[111] See, for example, Roger A. Pielke, Jr. and Mary Downton, "Precipitation and Damaging Floods: Trends in the United States, 1932-97," *Bulletin of the American Meteorological Society*, October 15, 2000, pp. 3625-3637; and Roger Pielke, Jr., "Future Economic Damage from Tropical Cyclones: Sensitivities to Societal and Climate Changes," *Philosophical Transactions of the Royal Society A.*, 2007 (published online, doi: 10.1098/rsta.2007.2086).

INDEX

D

G

H

I

J

K

L

M

N

O

P

Q

R

S

T